D1713850

Cognitive Processes in Mathematics

KEELE COGNITION SEMINARS

1. *Cognitive processes in mathematics*
   Edited by John A. Sloboda and Don Rogers

Cognitive Processes in Mathematics

Edited by

John A. Sloboda and Don Rogers
Department of Psychology
University of Keele

KEELE COGNITION SEMINARS: 1

CLARENDON PRESS · OXFORD
1987

Oxford University Press, Walton Street, Oxford OX2 6DP
Oxford New York Toronto
Delhi Bombay Calcutta Madras Karachi
Petaling Jaya Singapore Hong Kong Tokyo
Nairobi Dar es Salaam Cape Town
Melbourne Auckland
and associated companies in
Beirut Berlin Ibadan Nicosia

Oxford is a trade mark of Oxford University Press

Published in the United States
by Oxford University Press, New York

British Library Cataloguing in Publication Data
Cognitive processes in mathematics.—
(Keele cognition seminars; 1)
1. Mathematics—Study and teaching.
2. Cognition.
I. Sloboda, John A. II. Rogers, Don. III. Series.
510'.1'9 QA11
ISBN 0 19 852163 4

Library of Congress Cataloging in Publication Data
Cognitive processes in mathematics.
(Keele cognition seminars; 1)
Papers presented at a conference held at the University of Keele in the spring of 1985.
Includes bibliographies and indexes.
1. Mathematics—Study and teaching—Psychological aspects—Congresses. 2. Cognition—Congresses.
I. Sloboda, John A. II. Rogers, Don. III. Series.
QA11.A1C58 1986 370.15'6 86–16437
ISBN 0 19 852163 4

Printed in Great Britain by
Thomson Litho, East Kilbride, Scotland

Preface

DON ROGERS, ROBERT SIEGLER, and JOHN SLOBODA

This volume is the first in a series on current issues in cognition, which will appear every two years—the next is to be on social processes in cognition.

In choosing topics for this series, we are looking at issues which (a) are of fundamental importance, (b) are attracting interesting research, and (ideally) (c) involve the work of practitioners as well as theoretical workers. For these reasons mathematics represents a perfect starting place: the issues involved are not only important, but are interestingly diverse; the research being done in this area seems in some cases to be reaching a worthwhile consensus, while practitioners are able to show clearly that there are very many areas which still need a rigorous empirical and theoretical analysis.

The contributions published here are derived ultimately from papers given at a conference held at the University of Keele in the spring of 1985. We are grateful to the Nuffield Foundation for financial support for what turned out to be a most enjoyable and fruitful conference, and are only sorry that it has not proved possible to publish all the papers given there. In particular we regret that circumstances have prevented Richard Young from developing the extraordinarily interesting analysis in which he described how multiple 'agents' (rather like Selfridge and Neisser's 'demons') could seek partial matches in a problem and could thus enable children to cope with both familiar and unfamiliar problems.

One clear emphasis within the research presented at the conference was a focus on early competence within each area that was considered. Probably the most dramatic example of this trend was Starkey and Klein's paper on enumerative processes in infrahuman species, infants, and very young children. The paper reminds us of the biological basis of mathematical skills, a point that is easy to forget in an area with as much direct instruction as mathematical understanding. Other papers focused on areas where older children are just beginning the process of mastery. Hitch, Cundick, Haughey, Pugh, and Wright systematically investigated the counting procedures of pre-schoolers and early elementary school children. De Corte and Verschaffel investigated the usefulness of having first-graders retell the types of arithmetic story problems that they are just starting to learn. Resnick, Cauzinille-Marmèche, and Mathieu concentrated on children's reactions to the earliest algebra problems that they encounter. Even when adults were being considered, the emphasis was frequently on their initial unskilled performance. Smith's analysis of undergraduates' biases and errors in statistical judgements is an example of this emphasis on adults' early, unskilled performance.

Another focus of the papers was on how people represent mathematical information. Many of the enduring representational issues are currently being examined in the context of mathematical understanding; the results have great resonance because they inform our understanding both of people's knowledge of mathematics and of the representational issues in general. Greer probed the relations between imaginal and propositional understandings in children's approaches to arithmetic word problems. Young's paper and that of Resnick, Cauzinille-Marmèche, and Mathieu examined the relation between semantic and syntactic knowledge of mathematics. Todd, Barber, and Jones investigated whether representation of numbers was predominantly analogue or predominantly digital. Perhaps the most popular representational issue concerned the roles of associative and rule-based knowledge. Campbell and Graham examined the role of purely associative knowledge in multiplication, and Siegler looked at how rules and associations might work together in simple addition and subtraction.

The papers presented at the conference also pointed towards directions in which future research may proceed. One direction is toward more precise specification of change processes. We seem to be doing quite well in specifying particular states of knowledge, but need to make progress in accounting for how changes occur. Another, related direction concerns integrating early and advanced states of knowledge. One particularly nice example of how this can be done is demonstrated in the pair of papers by Graham and by Campbell. Working within the same network-model framework, Graham presented a model of early understanding of multiplication, Campbell a model of advanced understanding. A third direction for future research that these papers point to involves examination of more complex tasks. The majority of papers at the conference concerned such comparatively simple tasks as counting, magnitude comparison, and simple arithmetic, but a substantial minority concerned such complex tasks as algebra and inferential statistics. A final need felt by many participants was for higher level theories of mathematical understanding. Young's descriptions of how multiple agents engaging in partial matches might allow children to solve both familiar and unfamiliar problems is an exciting step in this direction. Rules and the kind of problem-specific associations described by Graham and Campbell, among others, will almost certainly play important roles in future theories.

Contents

Contributors

P. J. Barber, Department of Psychology, Birkbeck College, University of London, Malet Street, London WC1E 7HX, UK.

Jamie I. D. Campbell, Department of Psychology, Carnegie-Mellon University, Schenley Park, Pittsburgh, PA 15213, USA.

Evelyne Cauzinille-Marmèche, Laboratoire de Psychologie Genetique, Université René Descartes, Paris, France.

Jill Cundick, Department of Psychology, University of Manchester, Manchester M13 9PL, UK.

E. De Corte, Faculteit der Psychologie en Pedagogische Wetenschappen, Katholieke Universiteit Leuven, Vesaliusstraat 2, 3000 Leuven, Belgium.

David Jeffrey Graham, Department of Psychology, University of Waterloo, Waterloo, Ontario N26 3G1, Canada.

Brian Greer, Department of Psychology, Queen's University, Belfast BT7 1NN, Northern Ireland.

Maeve Haughey, Department of Psychology, University of Manchester, Manchester M13 9PL, UK.

Graham Hitch, Department of Psychology, University of Manchester, Manchester M13 9PL, UK.

D. Jones, Department of Psychology, Birkbeck College, University of London, Malet Street, London WC1E 7HX, UK.

Alice Klein, Department of Psychology, University of California, Santa Barbara, CA 93101, USA.

Jacques Mathieu, Laboratoire de Psychologie et Informatique, Université de Haute Normandie, France.

Rachel Pugh, Department of Psychology, University of Manchester, Manchester M13 9PL, UK.

Lauren B. Resnick, 824 Learning Research and Development Center, University of Pittsburgh, Pittsburgh, PA 15260, USA.

Don Rogers (editor), Department of Psychology, University of Keele, Keele, Staffordshire ST5 5BG, UK.

Robert S. Siegler, Department of Psychology, Carnegie-Mellon University, Schenley Park, Pittsburgh, PA 15213, USA.

Philip T. Smith, Department of Psychology, University of Reading, Whiteknights, Reading RG6 2AL, UK.

John Sloboda (editor), Department of Psychology, University of Keele, Keele, Staffordshire ST5 5BG, UK.

Prentice Starkey, Department of Psychology, University of California, Santa Barbara, CA 93101, USA.

R. R. Todd, Department of Psychology, Birkbeck College, University of London, Malet Street, London WC1E 7HX, UK.

L. Verschaffel, Faculteit der Psychologie en Pedagogische Wetenschappen, Katholieke Universiteit Leuven, Vesaliusstraat 2, 3000 Leuven, Belgium.

Hilary Wright, Department of Psychology, University of Manchester, Manchester M13 9PL, UK.

1

The origins and development of numerical cognition: a comparative analysis

ALICE KLEIN and PRENTICE STARKEY

The comparative approach to the study of animal and human intelligence has a long and distinguished history in psychology. Within the discipline of comparative psychology several different views of animal intelligence can be discerned (e.g. see Beer 1980). We will discuss two of these views in order to illustrate some of the contrasting assumptions that underlie comparative studies of intelligence.

One view is that of the traditional learning approach to behaviour. Although this view has generated a diverse body of research, ranging from Thorndike's (1898) investigations of problem-solving in animals to Skinner's (Skinner and Ferster 1957) applications of operant technology to a variety of species, all of the adherents share a fundamental assumption—continuity of learning processes in animals and man. Thus, species differences in task performance are attributed to *quantitative* differences in universal learning capacities. A corollary of this position is that the evolution of intelligent behaviour entails increments in learning capacities along phyletic lines from animals to man.

A contrasting view of animal intelligence, which has been proposed by Schneirla and his colleagues (e.g. Maier and Schneirla 1935), takes issue with both the continuity assumption and the quantitative differences assumption of the traditional learning approach. Instead, this view assumes discontinuity of learning processes in phylogeny, and it emphasizes *qualitative* as well as quantitative differences in intelligent behaviour across species including man.

Central to Schneirla's approach is the concept of psychological levels in animal behaviour. Specifically, animals can be classified by levels of behavioural organization rather than simply by position in a phyletic line. The expectation, then, is that an ordering of animals according to behavioural levels can follow non-linear phyletic relationships (i.e. animals of different evolutionary origins, such as birds and apes, can be at the same level) as well as linear ones. Gottlieb (1984) contends that such an ordering by levels of behavioural organization constitutes an 'anagenetic trend' based on the progressive evolution of adaptive behaviour. Anagenetic trends in intelligence are indexed by two principal features: increases in ontogenetic plasticity and behavioural versatility. Contrary to the traditional learning view, it is assumed that these features are manifested in qualitatively different forms of behavioural organization. Thus, in the behavioural-levels view, the evolution of intelligence is marked by a series of advances (not necessarily along phyletic lines) in the pattern of adaptive behaviours exhibited by animals and man.

The discussion of these contrasting views of animal intelligence raises two issues that bear on a method for the comparative study of cognition. The first issue concerns the relative importance of similarities and differences in comparisons between animal and human cognition. Recall that the traditional learning view focused on similarities and hence continuity in learning across species, whereas Schneirla emphasized differences and discontinuity in his view of levels of behavioural organization. In our comparative analysis, we follow a principle which has been advocated by Werner (1948), among others, that a comparative study must consider both similarities and differences. In order to apply this comparative principle, however, it is necessary to establish criteria for evaluating particular types of behaviours across species. One of the objectives of our analysis, then, will be to specify formal and psychological criteria for making phylogenetic comparisons of numerical behaviours.

The second issue follows from the assumption that there are qualitative, not just quantitative, differences between animal and human cognition. This assumption raises the issue of whether the cognitive systems of different species can be ranked according to increasing levels of intellectual capacity. Schneirla proposed that species should be classified by levels of behavioural organization which are ordered according to advances in the relative plasticity of their overall adaptive patterns (e.g. greater plasticity is indexed by less stereotypy and context-boundedness in the learning capacity of a given species). Although this proposal provides a metric of plasticity for ordering qualitatively different levels of intellectual capacity, it also has a major limitation.

Rozin (1976) has argued that plasticity is a fundamental component of intelligence, but it is not the only component: 'Consider an animal with an enormously complicated built-in program to handle one specific problem. It might greatly advance in power (intelligence) if it could apply that sophistication to another problem. . . . However, from the point of view of intelligence, the new program being accessed would be at least as interesting as the process of accessing it' (p. 258). Thus, in this view the complexity of the animal's knowledge (i.e. programme) as well as its plasticity are important considerations in the comparative study of cognition.

A further problem arises if complexity of knowledge is used as the basis for ordering species by their levels of intellectual capacity. Consider the example of a species that would be ranked high with regard to its spatial knowledge, but low with regard to its social knowledge. Which type of knowledge, spatial or social, best reflects the level of intellectual capacity of this species? Such a question is spurious in the absence of formal or psychological grounds for choosing one type of knowledge over another as the basis for ranking species by levels of intellectual capacity. We take the position that there are multiple domains or differentiated structures of knowledge (see Fodor 1983; Keil 1981; Langer, in press *a*; and Turiel and Davidson 1985 for explications of this position), and therefore, comparative studies of cognition must be made within each domain of knowledge. We have selected the domain of number for our analysis.

A comparative method for the study of numerical cognition

Our intent in this chapter is to analyse the origins and development of numerical cognition from a comparative perspective. In order to clarify the focus of our analysis, we make a distinction between two senses of the phrase, 'origins of numerical cognition'. One sense, which is outside the scope of this paper, denotes the point in man's evolutionary history or phyletic lineage at which numerical knowledge was first evident. The other sense refers to the primitive (elementary) levels of numerical knowledge found in human ontogeny and in phylogeny. It is this latter sense that our analysis addresses.

Specifying the primitive level of knowledge is fundamental to a developmental theory of number because the composition of the primitive level constrains the possible courses of ontogeny. But a developmental theory of number must also take into account the mature level (i.e. the extent) of knowledge that is attained in ontogeny, and the developmental mechanisms responsible for this attainment. If one examines only levels of knowledge within a single species such as man, however, it is very difficult to determine which primitive conditions are necessary and sufficient for the development of more advanced levels. We propose that these conditions can be specified, at least partly, through the use of a comparative method.

The comparative method that we use to examine the development of numerical cognition comprises two complementary analyses. One analysis is *synchronic* and the other is *diachronic.* A synchronic analysis is conducted at a particular level of numerical knowledge. It entails a description of the formal and psychological properties of the level of numerical behaviour exhibited by a species, regardless of the level of knowledge in other species. By contrast, a diachronic analysis is concerned with establishing developmental relations among levels of numerical knowledge in phylogeny. It is this type of analysis which can reveal an ordering of species from primitive to more advanced levels of knowledge.

In our analyses of numerical knowledge in human ontogeny and in phylogeny, we rely primarily on two comparative measures: *primitives* and *extents* of knowledge. These measures provide a basis for comparing the similarities and differences in numerical knowledge across species. We hypothesize that there are cross-species similarities in respect to some primitives, but there are differences in respect to the extents of numerical knowledge. In order to establish phylogenetic parallels in numerical knowledge, however, a further distinction must be made between material and formal differences. Langer (in press) has argued that it is not sufficient for comparative analyses to recognize material differences (e.g. species-specific differences in the content or range of application of a type of cognition to different environments) in cognitions that are similar on formal grounds; formal differences must be taken into account as well. One strategy for exposing formal differences in phylogeny is to compare the organization of knowledge (e.g. extents of ontogenetic development within domains) across species. For example, evidence from a study by Antinucci (1982) suggests that the sensorimotor development of logical and physical knowledge differs between non-human and human primates. Thus, the comparative method that we use to examine cross-species data on numerical behaviour will consider phylogenetic parallels as well as material and formal differences in numerical knowledge. In the next section we specify formal and psychological criteria for comparing types of numerical knowledge in human ontogeny and in phylogeny.

Types of numerical knowledge

In order to examine evidence for primitive levels of numerical knowledge in human ontogeny and in phylogeny, it is necessary to provide a framework of formal and psychological criteria for deciding whether some particular type of behaviour is truly numerical in nature, and if so, what numerical knowledge can be attributed to organisms which manifest this behaviour. In our view, a behaviour cannot be said to be numerical unless it implies knowledge of *one-to-one correspondence*, a foundational concept of mathematics (e.g. Frege 1950). As such, it can be used, for example, in establishing cardinal and ordinal relations among sets.

One-to-one correspondence, like much of mathematics, can be applied to problems in the physical world. One application is its use in establishing cardinal and ordinal relations among collections of objects. For example, in order to sort two collections (sets) of objects, set A and set B, on the basis of their cardinality, it is first necessary to establish whether the sets are equinumerous and hence identical in their cardinality. If every object in A can be paired with (mapped into) one and only one object in B, And if every object in B can be paired with one and only one object in A, then A and B are equinumerous. If this pairing is not possible, A and B are not equivalent. One-to-one correspondence can also be used to order non-equivalent sets with respect to their cardinal values. This can be done in the following way: if set A is equinumerous to a subset of B, but B is not equinumerous to a subset of A, then the cardinality of A is less than that of B. From a psychological standpoint, knowledge of one-to-one correspondence can be demonstrated only if the organism can represent collections of objects in a way that preserves the *discreteness* of the objects comprising the collections. A failure to represent collections in this way would preclude the use of one-to-one correspondence because one-to-one correspondence computations are performed on sets of discrete elements (objects).

One type of process, which is not truly numerical in nature because it does not entail knowledge of one-to-one correspondence, has been variously labelled as global quantification, numerical estimation (Klahr and Wallace 1976), and numerousness perception (Davis, Albert, and Barron 1985). Even so, the process of *global quantification* can be used to make impressively accurate discriminations between collections containing different numbers of objects. For example, a collection containing few objects is usually shorter or covers less surface area than one containing many objects. Thus, global quantitative features, such as the total surface areas or lengths of collections, rather than correspondences between discrete objects, can be used to discriminate between some collections. Particular quantitative features are sometimes highly correlated with the relative number of objects in two collections, but these features can sometimes be misleading indices of relative number. A longer row, for example, may contain fewer objects which are spread further apart than the objects in a shorter, more numerous row. Global quantification is not a primitive or precursory form of numerical knowledge, because, without a concept of one-to-one correspondence, discrete quantity can not be (formally) derived from global quantity (see Benacerraf and Putnam 1964).

The most elementary type of numerical knowledge is *enumeration.* Enumerative processes are numerical because one-to-one correspondence is psychologically instantiated in them. They are used to represent the number of objects in a collection and to compare collections in order to establish, for example, whether two collections are equivalent in number or whether the number of objects in a collection has changed from one time to another. We wish to draw distinctions among three enumerative processes: *numerosity perception, correspondence construction,* and *counting.*

One enumerative process has been variously identified in the literature as numerosity perception, apprehension of number (Fernberger 1921), or subitizing (Kaufman, Lord, Reese, and Volkmann 1949). This perceptual process constructs representations of small collections of objects which preserve the discreteness of the objects comprising the collections. The representations thus constructed can in turn be compared with one another to establish whether the collections are equal in number. From a psychological standpoint, numerosity perception is more elementary than other enumerative processes because it is a perceptual process.

A second enumerative process is correspondence construction. This psychological process maps every member of a collection of objects onto one and only one member of another collection of objects. A rather sophisticated use of this process is evident in the following account of a livestock census taken in a pre-colonial African state which had a taboo against counting animals:

On certain market days, a crier would be sent by the chief priest of the powerful spirit of the sacred river, to announce that the spirit threatened disaster to crops and livestock unless the people did as he bade. Every man and woman was to bring to the palace a cowrie shell for each animal he owned, and to deposit the shells in separate piles for sheep, goats, and cattle. First he must touch the animal with the cowrie, to transfer the danger from the animal to the shell. The king contributed an equal number of cowries, and retained a pebble for each shell he contributed. Thus the royal bureaucracy secured an accurate count of livestock, by kind and village, to be used as a basis for fiscal computations (Zaslavsky 1973, pp. 52–3).

A third process, counting, is also an enumerative process which entails knowledge of one-to-one correspondence. In contrast with the process of correspondence construction, counting maps a list of conventional (or idiosyncratic) number names onto a collection of objects. A unique number name is paired with each object in a collection, and the final number name that is used stands for the cardinal value of the collection.

We have used the presence of knowledge of one-to-one correspondence as a criterion for distinguishing numerical from non-numerical processes. The most elementary type of numerical process is enumeration. From a mathematical standpoint, processes that entail knowledge of a *number system* (i.e. an arithmetic) are more advanced than enumerative processes. A fully developed arithmetic system comprises knowledge of arithmetic operations (e.g. addition), mathematically necessary laws (e.g. commutativity), and the algorithmic use of enumerative processes in accordance with these operations and laws. In view of the formal and psychological criteria that we have proposed, we will next evaluate the existing data on numerical knowledge in animals and man.

Numerical knowledge in phylogeny

Do animals, apart from man, possess any knowledge of number, however elementary or contextually bound it might prove to be? The idea that other species might have elementary knowledge of number has arisen several times in the history of comparative psychology, and it has led to a variety of empirical studies. Our aim is not to offer a comprehensive review of all of the possibly relevant studies (see Davis and Memmott 1982 for a recent review of this type). Instead, we will critically discuss what we consider to be the strongest evidence to date for the existence of this domain of knowledge in animals. Most of this evidence is concentrated in a few species representing three animal groups: the class *Aves* (especially birds in the crow, parrot, and pigeon families), the class *Mammalia* (primarily mice and rats), and the order *primates* (primarily rhesus monkeys and chimpanzees).

There is an abundance of anecdotes about animals' numerical feats. Some have been debunked; for example, von Osten's clever horse, Hans, who reputedly mastered counting and decimal arithmetic at the height of his career (Watson 1914). But some anecdotes remain. Consider the following story about a 'ciphering' crow:

> A squire was determined to shoot a crow which made its nest in the watch-tower of his estate. Repeatedly he had tried to surprise the bird, but in vain: at the approach of man the crow would leave its nest. From a distant tree it would watchfully wait until the man had left the tower and then return to its nest. One day the squire hit upon a ruse: two men entered the tower, one remained within, the other came out and went on. But the bird was not deceived: it kept away until the man within came out. The experiment was repeated on the succeeding days with two, three, then four men, yet without success. Finally, five men were sent: as before, all entered the tower, and one remained while the other four came out and went away. Here the crow lost count. Unable to distinguish between four and five it promptly returned to its nest (Dantzig 1959, p. 3).

It is unfortunate for comparative psychologists (as well as for the crow) that the squire did not attempt a replication experiment. To this day, we have no confirmed evidence that crows are capable of such feats. Is there rigorous evidence, however, for the existence of any numerical knowledge in birds?

Birds

Koehler (1951) provided evidence which strongly suggests that a variety of bird species (e.g. the jackdaw, *Corvus monedula*, and the budgerigar, *Melopsittacus undulatus*) can perceive the numerosity of small collections of objects. As we will see from the evidence that follows, birds meet our criterion for attributing this process to their intelligence.

Some of Koehler's tasks did not produce findings which unequivocally demonstrated a perceptual knowledge of numerosity. For example, when given extensive training birds eventually learned to peck and eat a particular small number of seeds. They could take, say, exactly three seeds both when seeds were dispensed one-by-one and when a large pile of seeds was provided. Intriguing though this finding may be, it is difficult to use it to draw a conclusion about knowledge of numerosity. At issue is whether a representation of numerosity is guiding the animals' behaviour. It is certainly possible that Koehler's birds were representing the numerosities of collections of seeds or perhaps the numbers of pecks produced in pecking sequences. Alternatively, it is possible that the birds had simply learned to execute a single behavioural unit which behaviourally (i.e. to the outside observer) merely appears to comprise a sequence of three discrete elements or pecking behaviours. Representationally, the behavioural unit is not necessarily differentiated into three discrete elements; it could be a single complex action.

This latter alternative could also be used to explain the animals' performance on a match-to-sample task. On this task, birds eventually learned to choose an array that contained the same small number of items as a sample array. In this task, as in the preceeding one, these animals might have learned to execute unitary, undifferentiated behaviours which were afforded by arrays of particular numerosities. Just as stair-climbing is afforded by stairs, a two-peck sequence can be afforded by an array of two items. If the array were to be scanned systematically such that an encounter with a new item would elicit a (virtual) peck, birds could execute and match complex, unitary behaviours in the absence of numerical representations.

Theoretically, birds could do this, but they do not. Koehler found that his birds could learn to open a series of boxes which contained either zero, one, or two items of food. These birds learned to stop eating when a particular number of food items (e.g. three) had been found. Since the birds were able repeatedly to interrupt their pecking and eating behaviour in order to open the next box in the series, they were not simply executing a complex but undifferentiated behavioural unit. They could exit from and re-enter the behavioural unit at discrete points. Hence, after much training, the birds were utilizing a representation of a small number of discrete elements. The largest number they could cope with varied across family or species—five or six in pigeons, six in jackdaws, and seven in ravens and parrots.

Koehler's finding meets our criterion of acknowledging that some species of birds can perceive small numerosities. Birds can encode a collection in a way that preserves the discreteness and numerosity of the objects comprising the collection. Piaget (1971) has also interpreted Koehler's data as evidence for the existence of perceptual knowledge of number in birds. Piaget is correct in pointing out, however, that these data provide no evidence for the existence of a primitive number system in birds. The existence of system properties cannot be inferred from the existence of numerosities. It is possible that birds' representations of numerosities are isolated values which are neither ordered nor iterated in a number system.

Lower mammals

The rat has been used in a large majority of the studies of numerical knowledge in lower mammals. Most of these experiments, however, have been poorly constructed from the standpoint of testing for a sensitivity to numerosity. For example, temporal and numerical properties of the discriminative stimuli were confounded in many experiments (see Hobson and Newman 1981 for a review). Sometimes this confounding was rather subtle, as in an experiment by Davis, Memmott, and Hurwitz (1975). The discriminative stimulus was a sequence of three electrical shocks. The interval between shocks comprising a sequence was randomly varied, but the duration of individual shocks was fixed. Thus, the animals may have simply timed individual shocks and summed across them until a total duration that was equivalent to the durations of three shocks had been experienced.

Another instance of confounding temporal with numerical properties arose in experiments in which rats and mice were taught to make a response a fixed number of times (e.g. Mechner 1958; Millenson, 1962). Specifically, the nature of the confound was that three responses spanned more time than two responses. In subsequent experiments, animals that had been trained to make a fixed number of responses were given the drug haloperidol. This drug slowed the rate of the animals' responses but did not affect the number of responses they produced (Laties 1972). This control, however, does not necessarily disconfound the temporal and numerical properties because the animals' internal clocks may have been slowed by the drug as well. The control also does not eliminate possible proprioceptive cues, such as muscle fatigue, for knowing when to terminate a sequence of responses. If these alternative explanations were to be ruled out by a properly controlled set of experiments, it would still be impossible to draw conclusions about numerical knowledge from data indicating that lower mammals can learn to produce a fixed number of responses. The interpretive problem would be the same as we encountered with Koehler's demonstration that birds can learn to peck and eat three seeds from a pile and then stop their responding: an undifferentiated representation can generate a set of seemingly discrete behaviours.

Unfortunately, no set of tasks as well designed and demanding from a numerical standpoint as Koehler's has been attempted with lower mammals. The most compelling evidence of the rat's sensitivity to quantity—but not numerosity—was obtained by Church and Meck (1984). This research found that rats, when given much training, learned to make one type of response, such as a lever press to sequences containing 'few' (e.g. two) sounds and a different response to sequences containing 'many' (e.g. four) sounds. Furthermore, they initially learned responses to tones sounding few or many times and then transferred these responses to lights flashing few or many times. We concur with Church and Meck's own characterization of these data as evidence for a discrimination that is not necessarily more precise than a few–many discrimination. Further research will be required to establish whether rats can make discriminations between one and two elements, between two and three, and so forth. Thus, at the present time we cannot conclude that rats or other lower mammals possess any numerical knowledge. They may possess nothing more than a process for discriminating among global quantities.

Non-human primates

Converging evidence from a number of studies has revealed that non-human primates, most notably chimpanzees, have some elementary numerical knowledge. Chimpanzees can perceive the numerosity of small collections of objects. They can also acquire knowledge of how to construct correspondences between collections of objects. Consider first the studies of numerosity perception in rhesus monkeys and chimpanzees.

In Harlow's laboratory, Hicks (1956) trained adolescent rhesus monkeys (*Macaca mulatta*) to choose an array of three geometric figures over arrays of one, two, four, or five figures. The arrays were constructed such that neither colour, size, shape, nor arrangement of figures were confounded with numerosity. The animals received thousands of trials, but they did learn to select the arrays of three figures. This learning transferred to other arrays of three which were novel in configuration and other non-numerical properties.

Ferster and Hammer (1966) trained juvenile chimpanzees (*Pan troglodytes*) to make unique responses to collections containing small numbers of entities. One response, turning off a subset of a set of lights, was made to a collection of one object, and a different response, turning off a different subset of the set of lights, was made to a collection of two objects, and so forth. Responses were not chained together in an ordinal sequence, so a counting process was not required. Also, the number of lights that were turned off when a particular numerosity was perceived was not equal to the number of objects in the perceived collection, so the enumerative process, one-to-one correspondence between objects, was not required by the task. Ferster and Hammer found that the animals were able to respond correctly to novel collections of objects if the collections were of a small, familiar numerosity. (No test for transfer to unfamiliar numerosities was attempted.) Training required thousands of trials. When collections of four objects were introduced, over 20 000 trials were administered before performance rose to a level of five correct responses out of six trials. The large number of trials may have been a consequence of the undue complexity of the response the animals had to learn.

Hayes and Nissen (1971) found that their juvenile chimpanzee, Viki, could learn to choose sets of two or three objects when paired with sets of other numerosities. Like the crow in Dantzig's story, a limitation was found when she was given the task of discriminating between sets of four and five. Viki also learned to solve a match-to-sample task which required her to choose a collection from a set of two collections and to match it to a sample that was numerically equivalent. When she had to choose between two numerically adjacent collections (e.g. two versus three rather than two versus four or five), she was successful only in matching those of one, two, or three. The larger collections, four, five, and six, were beyond her ability. On a third task, Viki showed little ability to imitate auditory sequences of taps. When a human model tapped once on a surface, Viki also tapped once, but when the model tapped two or more times, she produced 'a rapid tattoo of taps for more-than-one' (Hayes and Nissen 1971, p. 77). Thus, the chimpanzee seemed only to make a global discrimination of few versus many when confronted with auditory sequences of elements rather than visual collections of elements.

Woodruff and Premack (1981) have also found that the chimpanzee can solve a numerical match-to-sample task. Four juvenile chimpanzees and an adult chimpanzee, Sarah, were tested. Only Sarah had received 'language' training; that is, training in the use of plastic symbols (see Premack 1984 for a discussion of this training). The animals' task was to choose a stimulus from a set of two stimuli and to match it to a numerically equivalent sample. If the sample was a collection of, for example, three cups, the animals' task was to choose between a collection of two spools and a collection of three spools. Collections contained one, two, three, or four objects. If the sample contained *N* objects, one potential choice also contained *N* objects and the other contained either one fewer or one more than *N* objects. Woodruff and Premack found that the juveniles responded randomly but that Sarah matched the collections by their numerosity.

The studies we have evaluated thus far indicate that some species of non-human primates can perceive the numerosities of small collections of objects. We next consider the available evidence on their ability to construct correspondences between collections of objects.

Premack (1976) has attempted to train chimpanzees to use correspondence construction to produce a collection that is numerically equal to another collection. One task was to place one and only one small cube onto each of the blocks comprising a collection of wooden blocks. This, unfortunately, was unnecessarily difficult in that there were more cubes than blocks, rather than an equal number of both. Even after much modelling and guidance by the experimenter, training was not wholly successful. The animals' performance was described as follows: 'Even after reaching the point when they were likely to carry out one-to-one correspondence, they still "spoiled" the occasional trial by dribbling out an extra counter (cube), removing a needed one, looking quizzically into the trainer's eyes, asking in effect, "Am I really done?"' (Premack 1976, p. 264). When the animals were next given other materials, such as egg cartons whose individual holes might provoke a correspondence with individual cubes, the animals appeared to make configurations based on 'aesthetic' principles rather than on one-to-one correspondence. Only Sarah, the 'language-trained' animal, learned to perform a one-to-one correspondence task. She did so only after an end-of-task marker had been introduced. The marker was a bell which she rang after she had finished pairing a cube with each of the blocks.

It was thus possible to train Sarah to use correspondence construction in one task. Even so, this training either did not persist or it failed to transfer. This limitation was evident in a subsequent experiment conducted in order to learn whether Sarah could conserve number and quantity (Woodruff, Premack, and Kennel 1978). Their intent was to present Sarah with a task that was similar in some (but not all) respects to one of Piaget's (1952) number conservation problems. The experiment, however, never proceeded to the point at which a collection of objects underwent a length transformation. Sarah failed to use number to make accurate same–different judgments on the initial collections. Thus, even though Sarah had previously learned to use correspondence construction to produce two equinumerous collections, she failed to use this knowledge to make same–different judgments about two collections that were already constructed.

The chimpanzee thus has a capacity to learn to use correspondence construction. The chimpanzee's learning, however, has thus far failed to transfer to a new experimental context, and we know of no reports that this enumerative process is used by chimpanzees in any situation in the wild. Even so, the existence of this capacity casts doubt on certain evolutionary hypotheses about the phyletic origins of knowledge of one-to-one correspondence. Parker and Gibson (1979) proposed that correspondence construction arose in *Homo habilis*, a direct ancestor of present-day man (*Homo sapiens*), as an adaptation to social food-sharing situations. In support of this hypothesis, they cite evidence that chimpanzees do not actively distribute food among group members (e.g. Isaac 1969). Gould (1979) has criticized their hypothesis because it follows from a panselectionist theory that all mental abilities are specific adaptations to specific situations. In evolutionary biology, it is now recognized that natural selection is not the only source of change. Accordingly, the panselectionist framework has given way to a pluralist framework. Thus, Premack's finding, taken together with Gould's criticism, suggests that the food-sharing hypothesis is in need of revision.

This completes our evaluation of the empirical findings on numerical knowledge in animals. Some non-human species clearly possess at least rudimentary knowledge of number. Of the three animal groups that have been studied, both birds and primates seem to possess some elementary numerical knowledge. Experimental efforts have thus far failed to demonstrate this type of knowledge in lower mammals. Even though the rat, for example, possesses knowledge of other types, such as elementary knowledge of Euclidean space (Cheng and Gallistel 1984), it does not seem to possess knowledge of number. By using its global quantification processes, the rat can only make gross discriminations of few versus many things. Birds and non-human primates, on the other hand, can perceive the numerosity of small collections. The most extensive numerical knowledge has been found in non-human primates. Chimpanzees who have been given special training can use correspondence construction to form two equinumerous collections of objects. We will next review some of the findings on the nature and development of numerical knowledge early in human ontogeny.

Numerical knowledge in human ontogeny

Mathematics is one of man's greatest intellectual achievements, but what is the point during human ontogeny at which our earliest knowledge of number originates, and what is its extent? In just the past half-decade we have discovered that numerical knowledge is present from a very early point in ontogeny. Knowledge entailed by numerosity perception and correspondence construction is present during infancy. Counting and rudimentary arithmetic algorithms emerge during the transitional period between infancy and early childhood, and arithmetic knowledge develops during early childhood. We will first consider the findings on numerosity perception and correspondence construction during infancy.

Infants

Human infants can perceive the particular numerosity of collections containing one, two, or three elements. This was shown by the following sets of experiments. It was found that infants can discriminate between two arrays of dots that contain different numbers of dots. Infants from birth to six months were habituated to arrays containing two to six dots. Sets of arrays were designed in a way that controlled for infants' potential use of cues such as array length, density, brightness, or configuration as a basis for discrimination. Babies dishabituated to an array containing a new number of dots if the number was less than or equal to four (e.g. Antell and Keating 1983; Starkey and Cooper 1980). It was also found that infants can detect numerical similarities and differences among small arrays even when the items comprising the arrays are heterogeneous in size, colour, configuration, and texture (e.g. Starkey, Spelke, and Gelman in press; Strauss and Curtis 1984). Numerosity is thus a salient property of collections of objects.

It has also been found that infants can detect numerical correspondences between collections of visible objects and sequences of sounds (Starkey, Spelke, and Gelman 1983). Infants, six to eight months of age, were presented with two visual displays placed side by side. One display was a slide of two objects and the other was a slide of three objects. While the slides were visible, either two or three drum-beats were sounded from a location midway between the displays. Several displays, which were heterogeneous with respect to a variety of non-numerical properties such as colour and configuration, were presented across trials. The dependent measure of interest was the duration of the infants' looking to the numerically equivalent displays and to the non-equivalent displays. Systematic looking preferences were observed in the infants. These preferences were contingent on the numerosity of the auditory stimuli. Infants preferred a visual display that corresponded in number to the accompanying auditory stimulus. This finding has been extended to displays of three-dimensional objects by Termine, Spelke, and Prather (unpublished), and, by making methodological modifications, Benenson, Moore, Peterson, Reznick, and Kagan (unpublished) have obtained an intermodal numerical preference for the non-corresponding display. Thus, infants can perceive the numerosity of a variety of collections, including auditory and visual collections.

Davis et al (1985) have argued that infants possess only global quantification processes which are capable of producing nothing more than gross discriminations of few versus many elements, and hence infants possess no knowledge that is truly numerical in nature. This argument, however, is not supported by the data. Infants can make discriminations between sets of one and two elements, between sets of two and three, and between sets of three and four, but not between sets of four, five, or six. These numerical differentiations are too precise and the intermodal numerical correspondence is too abstract for a global quantification process (Starkey, Spelke, and Gelman 1985). We thus argue that infants can perceive particular small numerosities, and hence that they possess an elementary type of numerical knowledge.

Recent research on infant cognition reveals that the ability to construct correspondences between two or more collections of objects emerges during infants' second year (Langer 1980, 1986; Sinclair, Stambak, Lézine, Rayna, and Verba 1982; Sugarman 1983). All of these studies employed a similar methodology, in that the testing procedures were non-verbal and the data base primarily consisted of infants' spontaneous manipulations of objects. To illustrate, Langer (1980) presented infants with sets of four to eight objects (e.g. blocks, rings) made out of malleable or non-malleable materials and simply encouraged them to play with the objects. Infants' interactions with the objects were recorded, action by action, in spatio–temporal order. These interactions were analysed in respect to the forms of infants' constructions (e.g. equivalence relations between sets of objects) and the processes by which their constructions were produced.

Infants constructed correspondences between two small collections of minimal complexity (i.e. two collections of two objects each) at age 12 months (Langer 1980; Sugarman 1983). For example, a 12-month-old simultaneously placed a yellow car on an orange spoon with one hand and placed a second yellow car on a second orange spoon with the other hand (Langer 1980, p. 348). Between 12 and 24 months, infants' correspondences developed in three respects. First, correspondence constructions were extended to multiple collections which were of minimal complexity (Sinclair, Stambak, Lézine, Rayna, and Verba 1982). Second, correspondences constructed between two collections were extended in size; that is, the collections contained three or four objects each (Langer, 1986). Finally, infants' correspondences were extended to incorporate a greater number of spatial relations which were increasingly abstract, from containment (12 months) to support (18 months) to linear (30 months) relations (Sugarman 1983). Thus we see that infants' use of correspondence construction contrasts sharply with the chimpanzee's both in its spontaneity and in its extent.

Toddlers

Numerical knowledge develops rapidly during the transitional period between infancy and early childhood. One hallmark of numerical development during this period is the emergence of counting (Gelman and Gallistel 1978). In their second year, children begin to use conventional number names (Gopnik 1981). Our observations and parents' reports indicate that counting initially emerges in narrow contexts such as those in which the child is climbing stairs or participating in some familiar game with a care-giver. Gelman and Gallistel (1978) found that two-year-old children accurately counted sets of two, and sometimes three, objects. In producing accurate counts, children reliably used an ordered list of number names and they placed these names into one-to-one correspondence with collections of objects. They thus exhibited knowledge of two fundamental principles of counting: a *one–one principle* that each object in a collection is to be paired with one and only one number name, and a *stable order principle* that each name is assigned to a permanent ordinal position in the list. Children of this age, however, did not exhibit knowledge of a third principle of counting, a *cardinal principle,* that the final number name used in a counting sequence refers to the cardinal value of the collection.

Gelman and Meck (1983) found that three-year-old children exhibited knowledge of the cardinal principle as well as the one–one and stable order principles if the children perceived another person counting and hence were not required to produce the counting by themselves. Gelman and Meck used an error detection procedure to establish that children discriminated counting errors (i.e. violations of any of the above three counting principles) from non-errors (i.e. correct, conventional counting) and pseudo-errors (i.e. correct but unconventional counting such as counts which began in the middle rather than at one end of a row of objects that were to be counted). Children were required to judge instances of a puppet's counting as 'okay' or 'wrong'. In one experiment, for example, the puppet counted a collection and reported either the final number name used in the count (i.e. the correct cardinal value) or some other number name (i.e. an erroneous cardinal value). Children detected errors of this kind, and they identified correct counts even when the numerosity of the collections that the puppet counted exceeded that of the collections the children could count by themselves. Briars and Siegler (1984) also studied young children's ability to detect counting errors and obtained similar results to those of Gelman and Meck. These data indicate that toddlers applied counting principles to large collections, and, therefore, their knowledge is not limited to small collections.

During this transitional period in human ontogeny, an elementary type of arithmetic algorithm also develops. One of the present authors (Starkey, unpublished) has found that two-year-old children computed the effects of addition and subtraction transformations on collections of objects. The experimental procedure was a variant of the search paradigm that has been used in several studies of cognition during infancy (e.g. Harris 1983). The child's task was to place a small collection of perceptually similar objects (balls) one at a time into a container. The container was constructed such that objects placed in it were screened from the child's view. Moreover, the container yielded only one object per reach by the child. This was accomplished by a false floor and trapdoor which operated such that only one object was actually in the main chamber of the container while the child searched. After the child had placed the objects in the container, the experimenter visibly added an object to or subtracted an object from the screened collection. The experimenter then instructed the child to remove all the objects from the container. Starkey found that two-year-olds exhibited some knowledge of a procedure for computing the effects of these transformations, but their knowledge was limited to small collections of objects. Children searched a correct number of times on the addition and subtraction problems when both the initial and the transformed collections contained three or fewer objects. Thus, toddlers as young as age two years modified their representation of a small collection in a way that was consistent with an addition or subtraction transformation that was performed on the collection. We will next consider some findings on the development of arithmetic knowledge during early childhood.

Young children

Evidence concerning the development of numerical knowledge during the first three years indicates that the enumerative processes of counting and correspondence construction emerge early in human ontogeny. Nevertheless, young children's understanding of these enumerative processes continues to develop as they form a number system during early childhood (Gréco 1962; Piaget 1952).

One consequence of this development is children's ability to use their enumerative processes, particularly counting, to compute solutions to simple arithmetic problems. Research has revealed that three- to five-year-old children employed, with varying degrees of accuracy, a diverse set of computational strategies to solve addition problems (Siegler and Robinson 1981). Moreover, children as young as age four years invented counting-based addition algorithms, such as counting on from the larger of two addends (Groen and Resnick 1977). Children's invention of arithmetic algorithms has been documented not only for a verbal counting system, but also for a body-part counting system that uses physical entities (body parts) rather than number names to represent numerical relations (Saxe 1982). Although there is considerable evidence that children develop counting alogrithms to facilitate arithmetic computation, little is known about their invention of computational procedures based on correspondence construction. We will now consider a study that compared children's use of these two enumerative processes in solving arithmetic problems (Klein 1984).

Klein examined the role of counting and correspondence construction in young children's arithmetic reasoning. The focus of this study was on developmental changes in how children make arithmetic inferences; that is, inferences about the outcome of addition and subtraction operations.

Pre-school children from ages four to six years participated in the study. The study comprised two experiments: a correspondence construction experiment and a counting experiment. The principal difference between the experiments was the enumerative process by which children formed their numerical representations of arithmetic problems. Children received a set of arithmetic problems in the context of a 'dinosaur game'. Each arithmetic problem was presented as a pair of linear collections of seven to ten objects (toy dinosaurs). The testing procedure for each problem consisted of an initial phase and a final phase.

The initial phase required the child to use a particular enumerative process to form a numerical representation of the arithmetic problem. In the correspondence construction experiment, the child was asked to look at two collections of objects which were arranged in one-to-one correspondence. The child was then asked to compare the collections and to give an explanation. In the counting experiment, the child was asked to count two collections of objects which were not arranged in one-to-one correspondence and to give the cardinal value of each of the collections. As in the correspondence construction experiment, the child was then asked to compare the collections and to give an explanation.

The final phase of the procedure assessed the child's ability to make arithmetic inferences. The experimenter performed an addition or subtraction operation on one or both collections and screened the final collections from the child's view. The screening procedure served to ensure that the child made an inference about the arithmetic operation and did *not* directly enumerate the final collections. The child was then asked to make a comparative judgment about the outcome of the arithmetic operation on the collections—'Do we have the same number of dinosaurs, or do you have more dinosaurs, or do I have more dinosaurs?'—and to give an explanation for this judgment.

Evidence about children's numerical representations of the arithmetic problems was obtained by examining their explanations in the initial phase of each experiment. These data revealed a similar developmental trend in both experiments: older children incorporated a fundamental property of the enumerative process into their numerical representations of the problems. Specifically, children in the correspondence construction experiment referred to the *correspondence relation* between objects in the two collections (e.g. 'Because this one is straight behind that one'), whereas children in the counting experiment referred to the *cardinal values* of the collections (e.g. 'Because I got ten and you got five'). Now consider the data on how children used their numerical representations of the problems to make arithmetic inferences.

The explanations that children gave for their judgments in the final phase of each experiment were analysed to determine their procedures for making arithmetic inferences. Of particular relevance to this analysis are children's explanations on inequality problems (i.e. problems in which an addition or subtraction was performed on one of two unequal collections) because these problems required an arithmetic computation.

The principal finding in both experiments was the development of children's use of computational procedures for making arithmetic inferences. Children's numerical representations of the problems, however, influenced the specific type of computational procedure that they employed. Let us examine the results of each experiment more closely.

In the correspondence construction experiment, there was a developmental shift in children's predominant type of explanation such that only the older children gave *configurational arithmetic* explanations which entailed the use of a computational procedure. In configurational arithmetic, the child performs a computation on a representation of the 'gaps' (i.e. the correspondence relations between objects present in one collection and objects missing in the other collection) in the initial collections of a problem. A child who correctly judged that he had more dinosaurs gave the following configurational arithmetic explanation: 'Because I had three more [referring to the 'gaps' in the experimenter's collection] in the beginning, and two left the waterhole to go to yours, so I have more' (R.T., age 5.11). Thus older children, who initially represented the correspondence relations between objects in the two collections, mentally manipulated these relations through a computational procedure in order to make their arithmetic inferences.

A similar pattern of results was obtained in the counting experiment. Children's explanations again revealed a developmental trend toward the use of a computational procedure. Note, however, that the older children in this experiment gave *numeral arithmetic* explanations. In numeral arithmetic, the child performs a computation on a representation of the cardinal values of the initial collections. A child who correctly judged that she had more dinosaurs gave the following numeral arithmetic explanation: 'I had ten and you had seven—that's three more—and you put two in yours, so I still have one more than you' (S.M., age 5.6). Thus, as in the correspondence construction experiment, children's initial representations of the problems influenced their computational procedures. Recall that older children in the counting experiment initially represented the cardinal values of the collections. Then they mentally manipulated these cardinal values through a numeral arithmetic procedure in order to make inferences about the outcome of arithmetic operations.

In summary, this study demonstrated that different enumerative processes, counting and correspondence construction, led young children to construct different numerical representations of arithmetic problems. Furthermore, children's numerical representations of the problems clearly had consequences for how they made arithmetic inferences. In both the correspondence construction experiment and the counting experiment, there was a significant developmental trend toward children's reliance on computational procedures to solve arithmetic problems. Nevertheless, children's numerical representations of the problems determined the particular type of computational procedure—numeral arithmetic or configurational arithmetic—that they employed to make arithmetic inferences.

Concluding remarks: ontogeny and phylogeny of numerical cognition

Our critical examination of the existing data on numerical behaviour in animals supports a conclusion which is, perhaps, counter-intuitive: species other than man, including some distantly related species, possess a primitive type of numerical knowledge. Birds and non-human primates, as well as human infants, can perceive particular small numerosities. Thus, we argue that these species employ an enumerative process in which the psychological instantiation of the numerical component, one-to-one correspondence, is perceptual.

This process of numerosity perception nonetheless differs across species in at least two respects. First, there are differences in the magnitude of the largest perceptible numerosity. Whereas birds can perceive the numerosity of collections of up to seven entities, chimpanzees are limited to collections of three of four, and human infants are limited to collections of three. Second, there are differences in the types of entities that can be enumerated by this process. Both birds and chimpanzees enumerate sets of visible entities. Chimpanzees, however, fail to enumerate sets of audible entities. (The relevant data are not available for birds.) By contrast, human infants enumerate sets of audible entities as well as sets of visible entities, and they can correlate these entities intermodally. These differences suggest that the enumerative process of numerosity perception may be part of the animal's system of visual perception, but in man it is part of a broader system. Thus, human infants can apply their perceptual enumerative process to a larger variety of enumerable entities.

There was only one non-human species, the chimpanzee, which exhibited an enumerative process more advanced than numerosity perception. Premack's 'language-trained' chimpanzee, Sarah, demonstrated some capacity to establish correspondence construction between collections of objects. Recall, however, that Sarah received both general 'language-training' and specific one-to-one correspondence training before she was able to manifest this enumerative process. Even so, Sarah did not transfer her correspondence training to a new experimental context in which she was required to use her one-to-one correspondence knowledge to make a same–different judgment about the initial collections in a number conservation task. Sarah's failure to transfer her correspondence training to a conservation context contrasts sharply with young children's success in transferring their 'trained' numerical knowledge in a variety of conservation contexts (see Beilin 1978 for an extensive review of this literature). Human infants' spontaneous production of correspondences that incorporate a variety of spatial relations (e.g. containment and support relations) provides a further contrast with Sarah's contextually bound one-to-one correspondence training. Although both the chimpanzee and the human infant and child possess knowledge of the enumerative process of correspondence construction, these species differ in regard to the range of contexts in which they can use this process. These cross-species differences in the extent of numerical knowledge support the view that there is discontinuity of numerical knowledge in phylogeny.

These differences also suggest an ordering of species with respect to the extent of their numerical knowledge. Lower mammals (rats and mice) have no apparent numerical knowledge and thus must be positioned at the lowest level. Since birds possess only the enumerative process of numerosity perception, they can be positioned at the next level. Chimpanzees have a capacity for the enumerative process of correspondence construction as well as numerosity perception, so they can be positioned at a higher level than birds. Finally, humans, who not only possess all three of the enumerative processes but also proceed to a more advanced level of numerical knowledge, the development of a number system and beyond, must occupy the highest position in this ranking. Note that this ordering of species by extent of numerical knowledge is an instance of a non-linear phyletic trend, in that birds are ranked above lower mammals.

Our comparative analysis of numerical cognition is instructive in specifying the primitive conditions which are necessary and sufficient for development of more advanced levels of knowledge. Based on the empirical evidence available at the present time, we conclude that a formally similar level of primitive numerical knowledge, numerosity perception, is present in birds, non-human primates, and human infants. These species differ, however, in respect to the extent of numerical knowledge they ultimately attain. For birds, numerosity perception marks the most advanced level of numerical knowledge. Chimpanzees manifest some capacity for one-to-one correspondence between objects; however, they require special training. Humans develop beyond the level of enumeration and construct a number system. These differences in the extent of numerical knowledge across species imply differences in formally similar types of knowledge as well. For instance, birds, chimpanzees, and human infants all possess a process of numerosity perception, but human infants can apply their process to a relatively broad set of enumerable entities. Another example is the finding that both chimpanzees and human infants can establish one-to-one correspondence between objects, but only human infants exhibit this process spontaneously and in a wide range of spatial contexts. Thus, the primitive knowledge entailed by numerosity perception is necessary for further numerical development, but is it not sufficient. Although this comparative analysis has not revealed what the *sufficient* conditions are, it has offered some intriguing clues. One clue is the relatively high degree of integration of the human infant's process of numerosity perception and the range of contexts in which infants and toddlers use correspondence construction. Another clue is the unique organization of the primitive level in humans, an organization comprising both perceptual and non-perceptual enumerative processes. Either or both of these clues may help answer the question of how advanced mathematical cognition is possible.

Acknowledgement

Preparation of this chapter was supported in part by NICHHD postdoctoral fellowship (2T32HD07181–06) to A. Klein.

References

Antell, S. E. and Keating, D. (1983). Perception of numerical invariance by neonates. *Child Development* **54,** 695–701.

Antinucci, F. (1982). Cognitive development in a comparative framework. In *La teoria di Jean Piaget* (ed. L. Camaioni), pp. 1–14. Barbera, Florence.

Beer, C. G. (1980). Perspectives on animal behavior comparisons. In *Comparative methods in psychology* (ed. M. H. Bornstein), pp. 17–64. Erlbaum, Hillsdale, NJ.

Beilin, H. (1978). Inducing conservation through training. In *Psychology of the 20th century: Vol. 7. Piaget and beyond* (ed. G. Steiner), pp. 260–89. Kindler, Zurich.

Benacerraf, P. and Putnam, H. (eds.) (1964). *Philosophy of mathematics.* Prentice-Hall, Englewood Cliffs, NJ.

Briars, D. and Siegler, R. S. (1984). A featural analysis of preschoolers' counting knowledge. *Developmental Psychology* **20,** 607–18.

Cheng, K. and Gallistel, C. R. (1984). Testing the geometric power of an animal's spatial representation. In *Animal cognition* (eds. H. L. Roitblat, T. G. Bever, and H. S. Terrace), pp. 409–23. Erlbaum, Hillsdale, NJ.

Church, R. M. and Meck, W. H. (1984). The numerical attribute of stimuli. In *Animal cognition* (eds. H. L. Roitblat, T. G. Bever, and H. S. Terrace), pp. 445–64. Erlbaum, Hillsdale, NJ.

Dantzig T. (1959). *Number: the language of science.* Free Press, New York.

Davis, H. and Memmott, J. (1982). Counting behaviour in animals: a critical evaluation. *Psychological Bulletin* **92,** 547–71.

Davis, H., Albert, M., and Barron, R. W. (1985). Detection of number or numerousness by human infants. *Science* **228,** 1222.

Davis, H., Memmott, J., and Hurwitz, M. B. (1975). Autocontingencies: a model for subtle behavioral control. *Journal of Experimental Psychology: General* **104,** 169–88.

Fernberger, S. W. (1921). A preliminary study of the range of visual apprehensions. *American Journal of Psychology* **32,** 121–33.

Ferster, C. B. and Hammer, C. E. (1966). Synthesizing the components of arithmetic behavior. In *Operant behavior: areas of research and application* (ed. W. K. Honig), pp. 634–76. Appleton–Century–Crofts, New York.

Fodor, J. A. (1983). *The modularity of mind.* MIT Press, Cambridge, MA.

Frege, G. (1950). *The foundations of arithmetic.* Oxford University Press, Oxford.

Gelman, R. and Gallistel, C. R. (1978). *The child's understanding of number.* Harvard University Press, Cambridge, MA.

Gelman, R. and Meck, E. (1983). Preschoolers' counting: principles before skill. *Cognition* **13,** 343–59.

Gopnik, A. (1981). Development of non-nominal expressions in 1- to 2-year-olds: why the first words aren't about things. In *Child language—an international perspective* (eds. P. S. Dale and D. Ingram), pp. 93–104. University Park Press, Baltimore, MD.

Gottlieb, G. (1984). Evolutionary trends and evolutionary origins: relevance to theory in comparative psychology. *Psychological Review* **91,** 448–56.

—— (1979). Panselectionist pitfalls in Parker & Gibson's model for the evolution of intelligence. *The Behavioral and Brain Sciences* **2,** 385–86.

Gould, S. J. (1979). Panselectionist pitfalls in Parker & Gibson's model for the evolution of intelligence. *The Behavioral and Brain Sciences* **2,** 385–86.

Gréco, P. (1962). Une recherche sur la commutativité de l'addition. In *Etudes d'épistémologie génétique: Vol. 13. Structures numériques élémentaires* (eds. P. Gréco and A. Morf), pp. 151–227. Presses Universitaires de France, Paris.

Groen, G. J. and Resnick, L. B. (1977). Can preschool children invent addition algorithms? *Journal of Educational Psychology* **69,** 645–52.

Harris, P. H. (1983). Infant cognition. In *Handbook of child psychology. Vol. 2* (ed. P. H. Mussen), pp. 689–782. Wiley, New York.

Hayes, K. J. and Nissen, C. H. (1971). Higher mental functions of a home-raised chimpanzee. In *Behavior of nonhuman primates. Vol. 4* (eds. A. M. Schrier and F. Stollnitz), pp. 59–115. Academic Press, New York.

Hicks, L. H. (1956). An analysis of number concept formation in the rhesus monkey. *Journal of Comparative and Physiological Psychology* **49,** 212–18.

Hobson, S. L. and Newman, F. (1981). Fixed-ratio-counting schedules. In *Quantitative analyses of behavior: Vol. 1, Discriminative properties of reinforcement schedules* (eds. M. L. Commons and J. A. Nevin), pp. 193–224. Ballinger, Cambridge, MA.

Isaac, G. (1969). The food-sharing behavior of protohuman hominids. *Scientific American* **238,** 90–109.

Kaufman, E. L., Lord, M. W., Reese, T. W., and Volkmann, J. (1949). The discrimination of visual number. *American Journal of Psychology* **62,** 498–525.

Keil, F. C. (1981). Constraints on knowledge and cognitive development. *Psychological Review* **88,** 197–227.

Klahr, D. and Wallace, J. G. (1976). *Cognitive development: an information processing view.* Erlbaum, Hillsdale, NJ.

Klein, A. (1984). *The early development of arithmetic reasoning: numerative activities and logical operations* Doctoral Dissertation, City University of New York. *Dissertation Abstracts International* **45,** 375B–76B.

Koehler, O. (1951). The ability of birds to 'count'. *The Bulletin of Animal Behavior* **9,** 41–5.

Langer, J. (1980). *The origins of logic: six to twelve months.* Academic Press, New York.

—— (1986). *The origins of logic: one to two years.* Academic Press, New York.

—— (in press). A note on the comparative psychology of mental development. In *Ontogeny and history* (ed. S. Strauss), Ablex, Norwood.

Laties, V. (1972). The modification of drug effects on behavior of external discriminative stimuli. *Journal of Pharmacology and Experimental Therapeutics* **183,** 1–13.

Maier, N. R. F. and Schneirla, T. C. (1935). *Principles of animal psychology*. Dover, New York.

Mechner, F. (1958). Probability relations within response sequences under ratio reinforcement. *Journal of the Experimental Analysis of Behavior* **1,** 109–21.

Millenson, J. R. (1962). Acquired counting behavior in mice maintained under two reinforcement procedures. *Animal Behavior* **10,** 171–3.

Parker, S. T. and Gibson, K. H. (1979). A developmental model for the evolution of language and intelligence in early hominids. *The Behavioral and Brain Sciences* **2,** 367–408.

Piaget, J. (1952). *The child's conception of number*. Routledge and Kegan Paul, London.

—— (1971). *Biology and knowledge*. University of Chicago Press, Chicago.

Premack, D. (1976). *Intelligence in ape and man*. Erlbaum, Hillsdale, NJ.

—— (1984). Possible general effects of language training on the chimpanzee. *Human Development* **27,** 268–81.

Rozin, P. (1976). The evolution of intelligence and access to the cognitive unconscious. In *Progress in psychobiology and physiological psychology. Vol. 6* (eds. J. M. Sprague and A. A. Epstein), pp. 245–78. Academic Press, New York.

Saxe, G. B. (1982). Culture and the development of numerical cognition: studies among the Oksapmin of Papau New Guinea. In *Children's logical and mathematical cognition: progress in cognitive developmental research* (ed. C. J. Brainerd). Springer-Verlag, New York.

Siegler, R. S. and Robinson, M. (1981). The development of numerical understandings. In *Advance in child development and behavior. Vol. 16* (eds. H. W. Reese and L. P. Lipsitt), pp. 242–312. Academic Press, Orlando, FA.

Sinclair, H., Stambak, M., Lézine, I., Rayna, S., and Verba, M. (1982). *Les bébés et les choses*. Presses Universitaires de France, Paris.

Skinner, B. F. and Ferster, C. B. (1957). *Schedules of reinforcement*. Appleton–Century–Crofts, New York.

Starkey, P. and Cooper, R. G., Jr. (1980). Perception of numbers by human infants. *Science* **210,** 1033–5.

Starkey, P., Spelke, E. S., and Gelman, R. (1983). Detection of intermodal numerical correspondences by human infants. *Science* **222,** 179–81.

—— (1985). Reply to Davis, Albert, and Barron. *Science* **228,** 1222.

—— (in press). Numerical abstraction by human infants. *Cognition*.

Strauss, M. S. and Curtis, L. E. (1984). Development of numerical concepts in infancy. In *The origins of cognitive skills* (ed. C. Sophian), pp. 131–55. Erlbaum, Hillsdale, NJ.

Sugarman, S. (1983). *Children's early thought: developments in classification*. Cambridge University Press, New York.

Thorndike, E. L. (1898). Animal intelligence: an experimental study of the associative processes in animals. *Psychological Review: Series of Monograph Supplements* **2,** issue 8.

Turiel, E. and Davidson, P. (1985). Heterogeneity, inconsistency and asynchrony in the development of cognitive structures. In *Stages and structure* (ed. I. Levin), pp. 106–43. Ablex, Norwood.

Watson, J. B. (1914). *Behaviorism: an introduction to comparative psychology*. Holt, Rinehart & Winston, New York.

Werner, H. (1948). *The comparative psychology of mental development* (2nd edn). International University Press, New York.

Woodruff, G. and Premack, D. (1981). Primitive mathematical concepts in the chimpanzee: proportionality and numerosity. *Nature* **293,** 568–70.

Woodruff, G., Premack, D., and Kennel, K. (1978). Conservation of liquid and solid quantity by the chimpanzee. *Science* **202,** 991–4.

Zaslavsky, C. (1973). *Africa counts: number and pattern in African culture.* Lawrence Hill & Company, Westport, CT.

2

Aspects of counting in children's arithmetic

GRAHAM HITCH, JILL CUNDICK, MAEVE HAUGHEY, RACHEL PUGH, and HILARY WRIGHT*

Introduction

It is widely appreciated that counting is an important feature of young children's arithmetical skills (e.g. see Gelman and Gallistel 1978; Ginsburg 1977; Starkey and Gelman 1982). For example, overt counting behaviour is often observed when children are required to find out the number of objects in a group. Children also use counting to determine changes in number when physical transformations such as combination or separation are performed on groups of objects. Such transformations correspond to the arithmetical operations of addition and subtraction. Indeed, when children are just learning to do simple numerical calculations, they sometimes count on their fingers, having first used them to create groups corresponding to the numbers (Fuson 1982). Eventually, however, most children learn to do simple calculations without the construction of such external perceptual referents, nor any overt, directly observable counting behaviour. In such cases *covert* counting has been inferred from the observation that times taken to perform simple sums are linearly related to the sizes of the operands (Groen and Parkman 1972; Svenson 1975).

More generally, it seems that the child learning elementary mathematics uses counting in an expanding range of increasingly sophisticated activities, from reciting the number sequence to mental calculation, and that counting moves from being an overt to a covert activity. If so, it is important to ask what psychological processes supply the evident *continuity* from one activity involving counting to another, and also what are the *discontinuities,* if any, between counting in different activities. Such similarities and differences may help us to understand what the child is required to achieve in building more complex arithmetical procedures from simpler ones.

* Maeve Haughey conducted her experiment as part of an Ulster Polytechnic course requirement during a temporary placement at Manchester University. Jill Cundick, Rachel Pugh, and Hilary Wright conducted their experiments whilst undergraduates at Manchester University.

We begin here by considering some of the psychological processes that might be involved in using the counting sequence to perform three common but rather different arithmetical activities: (i) 'ballistic' recitation of the numbers from 1 to 20, (ii) determining the number of items in a visual array, and (iii) using the 'counting-on' strategy to find the sum of two digits. Let us suppose for simplicity that all three activities involve the control of an internal pointer to a representation of the number series, such that the pointer can be made to move to the next number from its current position in the sequence.

We can see that ballistic counting would require very simple control of the pointer since it is necessary only to start at the beginning of the number series and repeatedly move 'next' until the terminal item has been selected. There is also a requirement to keep track of the current position of the pointer so that omissions and repetitions are avoided.

The situation would be more complex in the case of counting the items in a visual array. Here we must suppose that as well as the pointer which manages the number sequence there is another which selects items from the external visual array. The movements of these two pointers must be co-ordinated in order to maintain a strict one-to-one correspondence between items in the array and elements in the number series. Furthermore, there must also be some means of keeping track of the pointer to the array to ensure that each item is counted once and once only. This is of course in addition to the requirement to keep track of the pointer to the number sequence.

We turn finally to the case of using the counting-on strategy to add together two digits, according to which children count up from the larger digit. For example, adding 5 + 3 by counting-on would involve counting 'six, seven, *eight*'. Although very common, this is of course only one of the many possible counting strategies children have been found to use (e.g. see Svenson 1975). Counting-on can involve using the fingers to construct a physical array in order to keep track of the count, or perhaps imagining such an array. In such cases, the control processes would obviously be similar to those we have just considered and would involve co-ordinating the movements of two pointers, one to the array and one to the number series. Alternatively, rather than simply tallying against an array, a further numerical count (e.g. 'one, two, *three*') could be used to control the counting-on sequence ('six, seven, *eight*'). In this case, movements of the pointer to the number series are themselves counted. This type of 'double' counting would evidently involve more complex control processes than array counting.

Although we have omitted discussion of such additional operations as starting and stopping the count, the foregoing observations illustrate a general idea that is of some interest. This is that across a diverse range of arithmetical activities, counting is embedded in a range of different control structures which vary considerably in complexity. Perhaps the simplest possibility is that these activities all share access to a common procedure for counting. Such a procedure would evidently involve overt verbalization in younger children, becoming covert as they grow older. On this view we would characterize the child learning a new arithmetical activity as having to learn to 'hook up' the common procedure into a new control structure.

For practical reasons we were unable to investigate children's learning of new arithmetical skills. We opted instead to begin our research by simply comparing counting across different arithmetical tasks within an age-group. We were particularly interested in *covert* counting since this is very commonly used by school-aged children, and we concentrated our efforts on two very different activities, finding the number of items in a visual array and adding together a pair of integers. The chief experimental method was to examine the effects of dual-task interference on each activity, and, in particular, interference designed to disrupt subvocalization or 'inner speech'. If there is a common counting procedure which derives from overt speech, performance of both activities should be impaired when subvocalization is disrupted. While this would not be strong evidence for a common counting procedure, the opposite result of differential sensitivity to dual-task manipulations would raise strong objections to this hypothesis.

Experiment 1

In this experiment, carried out by Jill Cundick, we investigated the processes children use to determine the number of items in a visual array. The starting point was evidence from studies of working memory in adults and children showing that subvocal rehearsal processess can be suppressed by requiring irrelevant vocalization during the memory task (Baddeley, Thomson, and Buchanan 1975; Hitch and Halliday 1983). We reasoned that if finding the number of items in an array involves subvocal counting, then performance should be disrupted by concurrent articulation.

We were able to refine our prediction since previous research, summarized by Mandler and Shebo (1982), has suggested that there are actually two separate processes for determining numerosity depending upon array size. Judgments about arrays containing small numbers of items are usually rapid and accurate, and decision times do not vary greatly with the number of items in the array. Responses to larger arrays are slower and less accurate, and response times increase linearly with the number of items in the display. It is thought that small arrays are estimated by a process of 'subitizing', which differs from the slower process of serially counting larger arrays. There is considerable controversy over the interpretation of subitizing (e.g. see Mandler and Shebo 1982 for a useful discussion), but one very common view is that it involves the use of learned canonical number patterns. We therefore predicted that irrelevant vocalization would disrupt numerosity estimates for large arrays, where counting is necessary, but would leave the non-verbal process of subitizing small arrays unaffected.

The subjects were six boys and six girls drawn from a local school with a mean age of 8.7. On each trial they were shown a 12 cm × 12 cm square outline containing from 1 to 10 items. These displays were shown on the VDU of an Apple II microcomputer and were viewed from a distance of about 0.5 m. The items were all identical, and were block drawings of simple 'houses'. They were presented at randomly chosen intersections of an invisible cm grid dividing up the square. Taking care not to make any explicit reference to counting, the experimenter asked the child to decide 'how many houses there are' and to respond as quickly and accurately as possible on a numeric key-pad. The key-pad contained buttons labelled 1 to 10 and was placed immediately in front of the VDU. On each trial the visual array remained visible until after a response had been made.

Each child performed a total of 80 trials over two sessions, arranged so that the control and articulation conditions were blocked in a counter-balanced order. The size of array on any trial was randomly determined, subject to the constraint that each size occurred equally often overall. The locations of the items were selected at random for each trial. On the articulation trials the child was asked to continue repeating 'blah blah blah' until after a response had been made.

Figure 2.1 shows mean response times and per cent errors as a function of array size for both the control and concurrent articulation conditions. Response latencies approximate a linear function for larger arrays but appear to follow a shallower function for arrays of 1, 2, and perhaps 3 items. Thus it appears that children were indeed able to subitize the smaller arrays but had to rely on serial counting for the larger ones. The subitizing range seems to be somewhat smaller and less well marked than in previous studies, possibly reflecting the use of relatively complex stimuli here occupying a relatively large visual angle. The usual experimental method involves presenting arrays of dots. Most importantly, however, it can be seen that concurrent articulation had a clear disruptive effect on both speed and accuracy for responses to the larger arrays, but had no effect whatsoever on decisions about small arrays.

In order to conduct statistical tests, latencies for arrays of 1, 2, and 3 items were averaged to give a single estimate of subitizing for each child. Similarly, latencies for arrays of 7, 8, and 9 items were averaged to give a measure of counting. We had set these groupings up in advance to be within the ranges of the two processes and to avoid arrays where they might overlap. Table 2.1 shows mean latencies and mean error rates for these two ranges. An analysis of variance of the latency data confirmed that counted arrays took longer than subitized arrays ($F(1,23) = 42.8; p < 0.001$) and that there was a significant effect of concurrent articulation (F (1,23) = 9.89; $p < 0.01$). The latter can be seen from the table to be entirely restricted to arrays in the counting range, and this was confirmed by a highly significant two-way interaction (F (1,23) = 10.1; $p<0.001$). Errors mirror the latency

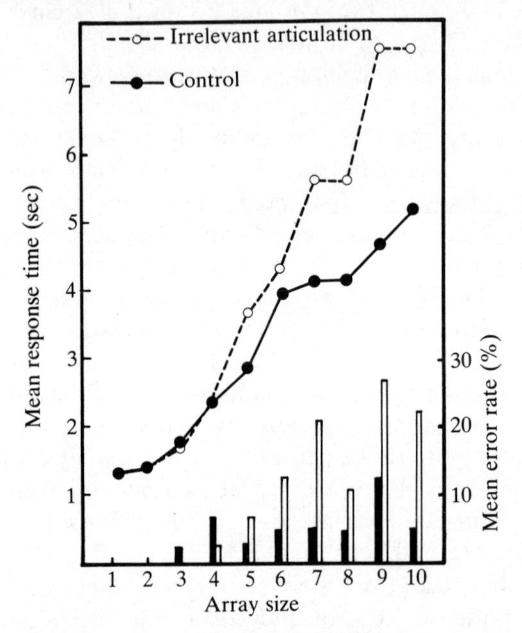

Fig. 2.1. Effect of concurrent articulation on estimates of numerosity as a function of array size.

Table 2.1. *Effects of irrelevant articulation on counted and subitized visual arrays in Experiment 1.*

	Control	Articulation
Mean latency (sec)		
1–3 item arrays	1.44	1.44
7–9 item arrays	4.31	6.27
Mean error rate (%)		
1–3 item arrays	0.7	0.0
7–9 item arrays	6.9	16.9

results, though there is evidently a floor effect within the subitizing range. Suppression did, however, significantly increase the number of errors in the counting range (t (11) = 3.10; $p<0.01$)*.

Our results are therefore consistent with the idea that the covert counting

* All t-tests are one-tailed unless stated otherwise.

of arrays involves subvocalization, and that subitizing small quantities does not. However, the evidence is not entirely convincing since it could be argued that the interference between counting and irrelevant vocalization reflects competition for general attentional resources involved in doing two things at once rather than competition specific to the articulatory system. A further experiment was therefore carried out to try to provide more decisive evidence on the involvement of subvocalization in counting arrays.

Experiment 2

In this study, carried out by Rachel Pugh, our method was to compare the interfering effects of concurrent finger-tapping with those of irrelevant articulation. We assumed that both these tasks would make similar demands on general attentional resources since they each involve the repeated production of a single, fairly simple response. However, only articulation involves the speech system. Thus if silently counting a visual array involves inner speech, it should suffer less interference from concurrent tapping than articulation. We opted to study only larger arrays of 6 or more items this time since we wished to focus on the counting process.

In a second part of the experiment, we used the same dual-task technique to investigate the processes involved in integer addition. If these involve subvocalization then they too should show greater interference from articulation than from tapping.

Eighteen schoolchildren with a mean age of 8.3 served as experimental subjects. In the array task, they were asked to count silently, and as quickly and accurately as possible, horizontal displays of from 6 to 11 items. The arrays were shown on a microcomputer-controlled VDU and comprised filled shapes chosen from one of four sets (matchstick man, heart, tree, rocket). The shapes were about 0.5 cm wide and between 0.5 and 0.8 cm high. They were equally spaced with a centre-to-centre separation of either 1.4 or 1.1 cm. The spacing between items and the horizontal position of the row on the VDU were varied randomly between trials to disrupt any tendency to use spatial location as a cue to numerosity. The display remained visible until the child responded by depressing the appropriate key on a specially adapted numeric key-pad. As before, the articulation task involved repeating the phrase 'blah, blah, blah' continuously until after a response was produced. The tapping task comprised tapping the table-top repeatedly with a finger of the non-preferred hand. In both interference conditions the children were required to emit responses at a fast rate throughout the trial. Whenever necessary they were reminded not to synchronize their responses with the counting sequence.

On each trial of the second task children were presented with a completed single-digit addition which was either correctly stated (e.g. $2 + 3 = 5$) or incorrect (e.g. 4 + 2 = 7). They were asked to decide as quickly and accurately as possible whether the problem should be given a 'tick' or a 'cross', and to press one of two labelled keys to indicate their decision. As before, the task was carried out with either concurrent articulation, tapping, or no concurrent task. Each condition was run in a block of 18 randomly ordered trials of which half comprised correctly stated sums and half were incorrect. The assignment of problems to conditions and the order of testing the conditions were counterbalanced across subjects. The order in which children performed the two tasks was also counterbalanced.

Table 2.2. *Array counting and integer addition performance in Experiment 3.*

	Control	Tap	Articulate
Array task			
Latency (sec)	5.42	5.45	6.59
Errors (%)	8.3	20.8	31.0
Addition task			
Latency (sec)	4.02	3.93	4.59
Errors (%)	6.8	6.8	13.0

The results for the array task are shown in Table 2.2, and suggest that articulation was more interfering than tapping. Individual comparisons on the error data showed a significant difference between the control and the pooled dual-task conditions ($t(17) = 4.82$, $p < 0.01$) and, more importantly, between articulation and tapping ($t(17) = 1.91$, $p < 0.05$). Similar tests on latencies just failed to reach significance ($p < 0.10$ in each case). Figure 2.2 shows mean times for correct responses as a function of array size. The plots are reasonably linear, consistent with the use of counting in all three experimental conditions. Analysis of variance confirmed a highly significant main effect of array size ($F(5, 85) = 29.4, p < 0.01$) with 99.3 per cent of the variability attributable to the linear trend. There was no interaction between interference and the effect of array size ($F < 1$). Thus, to sum up, the error data allow us to conclude that children's silent counting of visual arrays involves subvocalization. The latency data seem to suggest that children nevertheless show some ability to time-share between counting and irrelevant articulation.

Table 2.2 also shows the results for the arithmetic task. The absence of any interference from tapping was unexpected and suggests that arithmetic placed smaller demands on attention than the array task. Indeed, when we tried the tasks for ourselves we found that particular concentration was required to keep track within the visual array. Individual comparisons on the arithmetic data confirmed that articulation gave rise to a higher error rate than tapping ($t(17) = 2.65$, $p < 0.01$) and longer response times ($t(17) = 3.35$, $p < 0.01$). We conclude therefore that the silent counting used by children to carry out simple additions involves inner speech, just like their covert counting of visual arrays, and that the two tasks seem to involve different attentional loads.

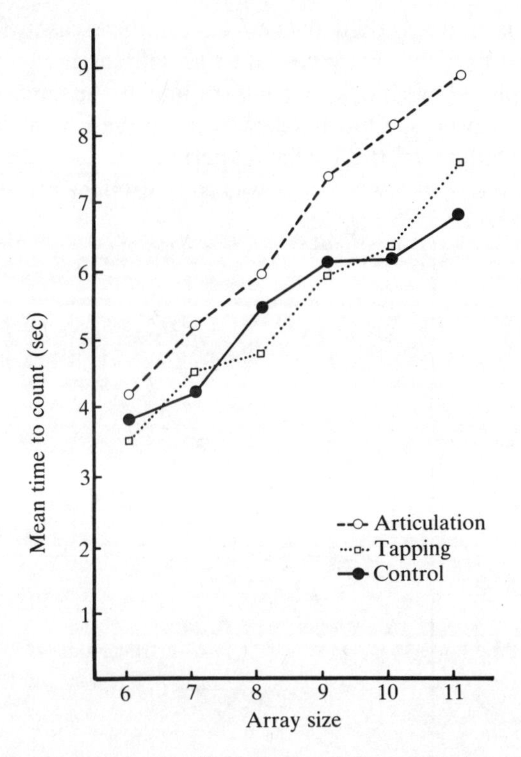

Fig. 2.2. Effects of concurrent articulation and finger-tapping on mean time to count arrays as a function of array size.

Unfortunately, the present experiment was not specifically designed to allow a check on whether the children were actually using counting algorithms to carry out the integer additions. We were fairly confident that this would be the case since, in an earlier study using the same experimental task, response latencies for nearly all the children studied were found to be consistent with use of a counting-on strategy (Hitch, Arnold, and Phillips 1983). In order to be more sure, however, an experiment carried out by Maeve Haughey investigated the effect of concurrent articulation on an addition task where we included an empirical check on the use of counting strategies.

Experiment 3

The subjects were 14 schoolchildren with an average age of 7.10. On each trial of the experiment they were shown an addition (e.g. 4 + 3 =) on the VDU of an Apple microcomputer. Children were asked to indicate the sum of the two numbers by pressing the appropriate key(s) on a standard numeric key-pad as quickly as they could without making 'silly' mistakes. Some of the sums had two-digit answers and, in these cases, unbeknown to the children, the first key-press was taken as defining the solution latency. Each child completed two blocks of trials containing the same 23 additions in different random orders. One block was performed with the concurrent articulation procedure of the earlier experiments, the other served as a control. The order of the two blocks was counterbalanced across subjects.

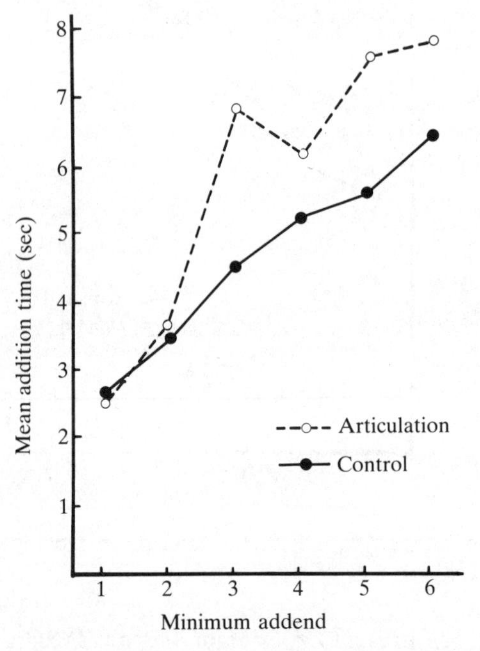

Fig. 2.3. Effect of concurrent articulation on time taken to add pairs of integers as a function of size of minimum addend.

The greatest effect of articulation was on accuracy. The mean error rate of 5.3 per cent in the control condition increased to 26.1 per cent with interference ($t(13) = 6.26, p < 0.01$). Mean solution latency also increased, from 4.64 sec to 5.71 sec ($t(13) = 2.01$, $p < 0.05$). Figure 2.3 is a plot of the mean latency of correct additions as a function of the 'minimum addend', the smaller of the two numbers to be added. This graph should be a straight line if children are using the counting-on strategy. Data from the control condition do appear to be closely linear, suggesting the use of this strategy. The graph is rather more noisy when children perform concurrent articulation. A two-way analysis of variance revealed a highly significant trend associated with addend size ($F(1, 65) = 66.1, p < 0.01$) accounting for 90.6 per cent of the variability. There was no support for the suggestion of an interaction between the effects of articulatory interference and array size.

To summarize, the results of our experiments are consistent with the hypothesis that children's covert counting in both integer addition and finding the number of items in a visual array involves subvocalization. Despite the numerous differences between these activities, they appear to require access to a common speech-based counting process. It would evidently be interesting to see whether these conclusions generalize to other arithmetical tasks since, as we shall suggest later, similarities in the components of complex tasks may be important in children's learning of new skills.

Given the importance of counting as a component of arithmetical skills which are acquired over a relatively long period of development, we wished to examine the possibility that it undergoes developmental change. We chose to focus on changes in the efficiency of counting in the period following initial learning of the counting sequence. The reasons for this will it is hoped become apparent in the closing discussion.

Experiment 4

In this study, carried out by Hilary Wright, we investigated developmental changes in the efficiency of counting the number of items in a visual array. We were primarily interested in changes in the *speed* of array counting. In particular, we wished to examine possible developmental changes in the relative speeds of counting arrays orally and silently. If covert counting involves subvocalization it may allow abbreviated articulation, with a consequent improvement in the speed of operations. Such an advantage might not be evident early in development, however, when the child is presumably just learning to internalize speech and thus expending attentional effort in actively inhibiting vocalization. We therefore anticipated the possibility of a cross-over in the relative speeds of silent and oral array counting with increasing age. Alternatively, the factor limiting performance may be the control processes involved in dealing with the *array,* rather than counting itself. In this case we would expect speeds of silent and oral counting to be the same.

We also included a measure of the speed with which children can recite the counting sequence from 1 to 20. According to our general hypothesis, oral recitation and oral array counting each require access to a common articulatory subroutine for stepping through the number sequence, but array counting involves a more complex control structure. Since children's speech rate increases with age (e.g. see Hitch and Halliday 1983), we expected that the speed of both tasks would show a broadly similar developmental increase. However, the extra control processes in array counting led us to expect that it would be the slower of the two tasks throughout development.

Children were shown a row of from 6 to 12 pictured objects using the method of display described in Experiment 2. There were four age-groups of subjects, each comprising 8 males and 8 females, with mean ages 6.1, 8.1, 10.0, and 21.4.

Each trial of the experiment began with a warning signal followed by presentation of a randomly selected array. The 'silent' and 'aloud' counting conditions were run in consecutive blocks in a counterbalanced order. When giving instructions to the younger children in the silent condition, the experimenter stressed that counting should be 'in your head, so I cannot hear you'. The experimenter stopped a timer in order to obtain response latencies. This was a straightforward matter when subjects counted aloud or when younger children counted silently since their lip movements could be easily monitored. Lip movements were less obvious for older subjects, however, and in this case the experimenter required that the final item be articulated aloud, as in '. . . five, six, *seven*'. These timing procedures are clearly not entirely satisfactory, but they are not thought to give rise to serious problems of interpretation here. Subjects were also timed in the ballistic counting task of reciting the number sequence from 1 to 20 both before and after the array counting task.

Mean error rates for the array counting task were reasonably low, the percentages for the four age-groups being 9.6, 5.1, 8.3, and 3.2 going from youngest to oldest. Figure 2.4 shows how mean counting time for correct responses varied as a function of the size of the array for the different groups and conditions. The functions appear to be linear, and it seems that the speed of counting decreases steadily across the entire range of age-groups. A three-way mixed analysis of variance confirmed a significant linear trend ($F(1,360) = 3769, p < 0.01$) accounting for 99.9 per cent of the variability associated with array size. The main effect of age was significant ($F(3,60) = 63.7, p < 0.01$) and there was also a significant interaction between age and array size ($F(18, 360) = 17.0, p < 0.01$).

The predicted interaction between age and mode of counting was highly significant ($F(3,60) = 4.51, p < 0.01$). One-tailed paired comparisons on data for the two extreme age-groups showed that 6-year-olds were significantly slower when counting silently ($t(15) = 3.84, p < 0.01$), while adults were slower when counting aloud ($t(15) = 1.77, p < 0.05$). Two-tailed tests of the differences for the two intermediate age-groups failed to approach

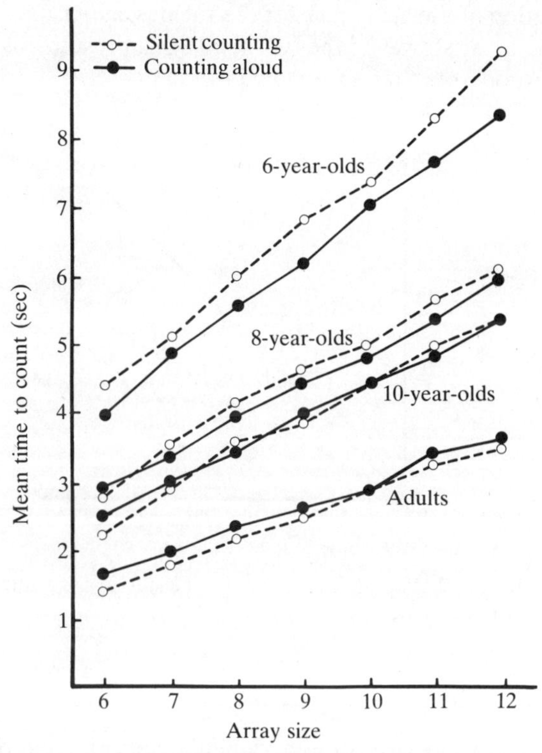

Fig. 2.4. Mean time to count arrays as a function of array size.

significance. Thus the relative efficiency of silent and oral array counting appears to reverse with increasing age.

The comparisons between silent and oral counting times are consistent with the idea that internalizing speech is at first effortful for the young child, and that the benefit of speed in processing occurs rather later in development. It is of some interest to note that we also tested 5-year-old children, but although they could count aloud quite competently, many of them could not follow the instruction to count silently. The youngest children included in the experiment were therefore not long past first learning to internalize counting. Turning to the adults, we were slightly surprised that they did not show a much greater advantage to silent counting, given our own subjective evaluations of the relative speeds of silent and overt articulation. The small observed difference suggests that the control processes involved in dealing with the array may have limited performance in older subjects rather than articulation itself.

A final aspect of the results is shown in Fig. 2.5 which plots the mean time to count one step in the ballistic counting task against the same measure for the oral condition of the array task for the various age-groups. As expected, array counting is the slower of the two at all ages, presumably reflecting the extra control processes that are involved.

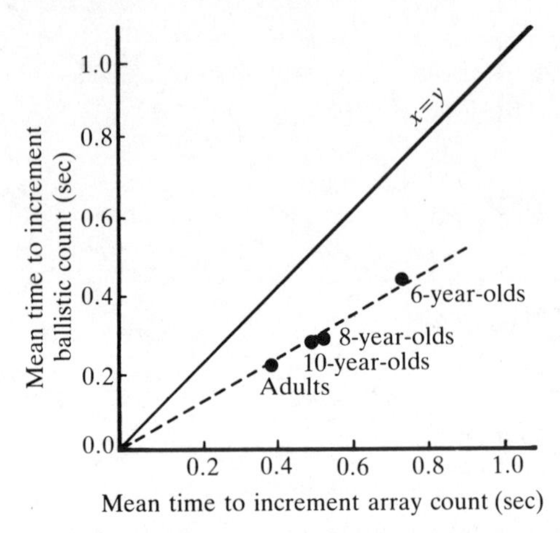

Fig. 2.5. Relationship between times to count one step in array and ballistic counting as a function of age.

General discussion

Although modest in scope, our experiments shed some interesting light on counting in children's arithmetic. They show that the speed with which both oral and covert counting are performed increases steadily throughout development, and that covert counting in two otherwise quite different arithmetical activities involves subvocalization. The overall pattern of results is consistent with the idea that covert counting develops through the internalization of oral counting and acts as a common procedure in different tasks. This implies that when children learn to perform a new arithmetical task, part of what they learn is to build a new control structure incorporating the common procedure. We speculate below that the speeding up of counting that is observed throughout development may be important for this building process to occur.

First, it is of some interest to note that the results of the experiments can be readily interpreted in terms of the theoretical concept of working memory (Baddeley and Hitch 1974). This model assumes that a range of cognitive tasks draw on the limited resources of a general attentional system (called the 'central executive') which controls specialist subsystems such as a verbal 'articulatory loop'. The model has been shown to be useful in analysing the development of children's short-term memory abilities (Hitch and Halliday 1983), and the proposition that there is a limited capacity, central 'workspace' for executing and monitoring control processes also appears in several approaches to general cognitive development (Case 1982; Halford and Wilson 1980; Pascual-Leone 1970). According to the working memory model, the counting procedure would involve the articulatory loop while the command structure which controls it would load the central executive. Thus, learning a new arithmetical procedure involving more complex control processes would increase the load on the executive. Indeed, the model makes the rather dramatic prediction that if the control requirements of a new task exceed the capacity of the executive, it should be impossible to learn.

Unfortunately, we know so little about how to measure the capacity of the executive that it would be difficult to carry out a direct test of the prediction. We do know, however, that practice in skilled activities increases the speed and fluency of their execution and decreases the demands they place upon attention. It may be, therefore, that the observed developmental speeding up in tasks involving counting is a practice effect, such that their control processes place an ever decreasing load on the central executive. If so, we can propose that such practice is *necessary* in order for tasks involving more complex control structures to be acquired. Thus we are suggesting that after first learning to recite the number sequence, the child must become fluent in its execution before this procedure can be incorporated within a sophisticated control structure such as that required for counting a collection of objects. With further practice the control processes of counting objects will themselves come to place a smaller burden on the central workspace so they in turn can become integrated into more complex procedures. Children might then, for example, become able to learn a procedure for counting imagined as well as perceived objects. For this to be the case, the model requires that the combined attentional demands of controlling imagery and counting would have to be within the capacity of the central executive.

It must of course be remembered that these theoretical ideas are speculative since our experiments have not been addressed to the learning process itself. It should also be noted that they are simplistic, and some of their more obvious limitations acknowledged. One is that we have guessed at the nature of the counting procedure in our assumptions about the movements of a pointer to a serial chain. In a thorough and stimulating analysis of the development of counting, Fuson, Richards, and Briars (1982) suggested that the structure of the counting sequence undergoes considerable modification as children grow older. The changes relate to such properties as whether there is random access to any location within the chain or whether it must always be entered at the beginning. Perhaps a more plausible idea than our suggestion of a 'common-core' counting procedure is that of a range of increasingly sophisticated counting routines, each of which is built from a simpler one. Such a complex hypothesis might, however, be very difficult to explore empirically. A second limitation concerns the obvious importance of a much wider range of factors in children's arithmetical learning than we have discussed here. In particular, we have not considered children's understanding of the concepts underlying different sorts of arithmetical activity, nor the possibility that conceptual difficulties may constrain new learning. In any complete analysis, it is clearly necessary to view children's arithmetical skills as comprising both control processes (skill) and special knowledge representations (understanding). Despite its incompleteness, however, the present approach appears to touch upon a potentially important source of difficulty when children are required to learn new arithmetical procedures. It may therefore be worth exploring further.

Acknowledgments

Thanks are due to Paul Arnold for useful discussion and collaboration in parts of this work, to Sebastian Halliday, Doug Herrman, and John Sloboda for comments on the manuscript, and to the schools in Greater Manchester that kindly co-operated in the research.

References

Baddeley, A. D. and Hitch, G. J. (1974). Working memory. In *The psychology of learning and motivation, vol. 8* (ed. G. H. Bower), Academic Press, New York.

Baddeley, A. D., Thomson, N., and Buchanan, M. (1975). Word length and the structure of short-term memory. *Journal of Verbal Learning and Verbal Behavior* **14,** 575–89.

Case, R. (1982). General developmental influences on the acquisition of elementary concepts and algorithms in arithmetic. In *Addition and subtraction: a cognitive perspective* (eds. T. P. Carpenter, J. M. Moser, and T. A. Romberg). Erlbaum, Hillsdale, NJ.

Fuson, K. C. (1982). An analysis of the counting-on solution procedure in addition. In *Addition and subtraction: a cognitive perspective* (eds. T.P. Carpenter, J. M. Moser, and T. A. Romberg). Erlbaum, Hillsdale, NJ.

Fuson, K. C., Richards, J., and Briars, D. J. (1982). The acquisition and elaboration of the number word sequence. In *Children's logical and mathematical cognition: progress cognitive development research* (ed. C. J. Brainerd). Springer–Verlag, New York.

Gelman, R. and Gallistel, C. R. (1978). *The child's understanding of number.* Harvard University Press, Cambridge, MA.

Ginsburg, H. (1977). *Children's arithmetic: the learning process.* Van Nostrand, New York.

Groen, G. J. and Parkman, J. M. (1972). A chronometric analysis of simple addition, *Psychological Review* **79,** 329–43.

Halford, G. S. and Wilson, W. H. (1980). A category theory approach to cognitive development. *Cognitive Psychology* **12,** 356–411.

Hitch, G. J. and Halliday, M. S. (1983). Working memory in children. *Philosophical Transactions of the Royal Society London, Series B* **302,** 325–40.

Hitch, G. J., Arnold, P., and Phillips, L. J. (1983). Counting processes in deaf children's arithmetic. *British Journal of Psychology* **74,** 429–37.

Mandler, G. and Shebo, B. J. (1982). Subitizing: an analysis of its component processes. *Journal of Experimental Psychology: General* **111,** 1–22.

Pascual-Leone, J. (1970). A mathematical model for the transition rule in Piaget's developmental stages. *Acta Psychologica* **32,** 301–45.

Starkey, P. and Gelman, R. (1982). The development of addition and subtraction abilities prior to formal schooling in arithmetic. *Addition and subtraction: a cognitive perspective* (eds. T. P. Carpenter, J. M. Moser, and T. A. Romberg). Erlbaum, Hillsdale, NJ.

Svenson, O. (1975). Analysis of time required by children for simple additions. *Acta Psychologica* **39,** 289–302.

3

Using retelling data to study young children's word-problem-solving

E. DE CORTE and L. VERSCHAFFEL*

Background and theoretical framework

Since 1979 we have been working on a research project in which an attempt is being made to acquire a better understanding of the development of young children's problem-solving skills and processes with respect to elementary arithmetic word problems, and of the influence of formal schooling on that development. Besides an orientation toward the construction of a theory of learning to solve arithmetic problems, and toward the development of a methodology for research on children's problem-solving, we are also concerned about the practical relevance of our work: it is hoped that our findings will be useful in view of the improvement of arithmetic instruction.

This research project is strongly influenced by the work of Greeno and associates in the late-1970s (Greeno 1978; Heller and Greeno 1978; Riley and Greeno 1978). These investigators introduced a new classification scheme for elementary addition and subtraction word problems based on the semantic relations underlying these problems. More specifically, they distinguish three main categories of word problems: change, combine, and compare problems. Change problems refer to dynamic situations in which some event changes the value of a quantity; e.g. 'Joe had 3 marbles; Tom gave him 5 more marbles; how many marbles does Joe have now?' Combine problems relate to static situations involving two amounts, that are considered either separately or in combination, as in the following example: 'Joe has 3 marbles; Tom has 5 marbles; how many marbles do they have altogether?' Compare problems involve two amounts that are compared and the difference between them; e.g. 'Joe has 3 marbles; Tom has 5 more marbles than Joe; how many marbles does Tom have?' Each of these three categories was further subdivided into distinct problem types depending on the identity of the unknown quantity; for the change and compare problems, further distinctions were made depending on the direction of the event (increase or decrease) or the relationship (more or less), respectively.

* Lieven Verschaffel is a Research Associate of the Belgian National Fund for Scientific Research.

Combining these three characteristics, Greeno and associates distinguished fourteen types (see Table 3.1).

Greeno and associates also developed a hypothetical model of elementary arithmetic word-problem-solving, in which semantic processing is considered to be the crucial component in skilled problem-solving (Greeno 1978; Heller and Greeno 1978; Riley and Greeno 1978). More recently the same authors have elaborated this model. For the three semantic categories of word problems included in their analysis (change, combine, and compare), Riley, Greeno, and Heller (1983) identified different levels of skill, each associated with a distinctive pattern of correct responses and errors on the problem types within the three categories. They developed models that simulate these different levels of performance, some of which are implemented in a running computer program. Briars and Larkin (1984) have proposed a similar computer-implemented model of children's solution processes with respect to elementary arithmetic word problems.

Based upon the early work of Greeno and associates on the one hand, and on the results of our own empirical investigations on the other, we have developed a competent problem-solving model of addition and subtraction word problems comprising five stages (De Corte and Verschaffel 1985*a*, 1985*b*; Verschaffel 1984):

(1) A complex, goal-oriented, text-processing activity occurs: starting from the verbal text, the pupil constructs a global, abstract, internal representation of the problem in terms of sets and set relations.

(2) On the basis of that representation, the problem-solver selects an appropriate formal arithmetic operation or an informal counting strategy to find the unknown element in the problem representation.

(3) The selected action or operation is executed.

(4) The problem-solver reactivates the initial problem representation, replaces the unknown element by the result of the action performed, and formulates the answer.

(5) Verification actions are performed to check the correctness of the solution found in the preceding stage.

As stated above, the first stage of the solution process is perceived as a goal-oriented, text-processing activity. The emerging problem representation is considered as the result of a complex interaction of bottom–up and top–down analysis; that is, the processing of the verbal input as well as the activity of the subject's cognitive schemata contribute to the construction of the problem representation. Two main categories of cognitive schemata are distinguished: (a) semantic schemata, representing the subject's knowledge about increasing and decreasing, combining, and comparing groups of objects (the change, combine, and compare schema, respectively), and (b) the word-problem schema, which involves knowledge of the structure of word problems, their role and intent in arithmetic instruction, and implicit

Table 3.1. *Types of word problems distinguished by Greeno and associates.*

Type	Example	Schema	Direction	Unknown
Change 1	Joe had 3 marbles; then Tom gave him 5 more marbles; how many marbles does Joe have now?	change	increase	result set
Change 2	Joe had 8 marbles; then he gave 5 marbles to Tom; how many marbles does Joe have now?	change	decrease	result set
Change 3	Joe had 3 marbles; then Tom gave him some more marbles; now Joe has 8 marbles; how many marbles did Tom give him?	change	increase	change set
Change 4	Joe had 8 marbles; then he gave some marbles to Tom; now Joe has 3 marbles; how many marbles did he give to Tom?	change	decrease	change set
Change 5	Joe had some marbles; then Tom gave him 5 more marbles; now Joe has 8 marbles; how many marbles did Joe have in the beginning?	change	increase	start set
Change 6	Joe had some marbles; then he gave 5 marbles to Tom; now Joe has 3 marbles; how many marbles did Joe have in the beginning?	change	decrease	start set
Combine 1	Joe has 3 marbles; Tom has 5 marbles; how many marbles do they have altogether?	combine	—	superset
Combine 2	Joe and Tom have 8 marbles altogether; Joe has 3 marbles; how many marbles does Tom have?	combine	—	subset
Compare 1	Joe has 8 marbles; Tom has 5 marbles; how many marbles does Joe have more than Tom?	compare	more	difference set
Compare 2	Joe has 8 marbles; Tom has 5 marbles; how many marbles does Tom have less than Joe?	compare	less	difference set
Compare 3	Joe has 3 marbles; Tom has 5 more marbles than Joe; how many marbles does Tom have?	compare	more	compared set
Compare 4	Joe has 8 marbles; Tom has 5 marbles less than Joe; how many marbles does Tom have?	compare	less	compared set
Compare 5	Joe has 8 marbles; he has 5 more marbles than Tom; how many marbles does Tom have?	compare	more	reference set
Compare 6	Joe has 3 marbles; he has 5 marbles less than Tom; how many marbles does Tom have?	compare	less	reference set

rules and assumptions underlying that particular text type (De Corte and Verschaffel 1985*a*, 1985*b*).

A longitudinal investigation contributed substantially to the development of our theoretical model. In that study we collected empirical data on first-graders' representations and solution processes with respect to elementary arithmetic word problems. Thirty children were individually interviewed three times during the school year: at the very beginning in September, in January, and at the end in June. Together with a series of Piagetian tasks, memory tasks, counting tasks, and numerical arithmetic problems, they were administered each time with eight word problems (see Table 3.2).

Table 3.2. *Overview of the eight word problems used in our study.*

Type	Problem*
Change 1	Pete had 3 (5) apples; Ann gave Pete 5 (7) more apples; how many apples does Pete have now?
Change 2	Pete has 6 (12) apples; he gave 2 (4) apples to Ann; how many apples does Pete have now?
Change 3	Pete had 3 (5) apples; Ann gave Pete some more apples; now Pete has 10 (14) apples; how many apples did Ann give to Pete?
Change 6	Pete had some apples; Pete gave 3 (5) apples to Ann; now Pete has 5 (7) apples; how many apples did Pete have in the beginning?
Combine 1	Pete has 3 (5) apples; Ann has 7 (9) apples; how many apples do Pete and Ann have altogether?
Combine 2	Pete has 3 (5) apples; Ann has also some apples; Pete and Ann have 9 (13) apples altogether; how many apples does Ann have?
Compare 1	Pete has 3 (5) apples; Ann has 8 (12) apples; how many apples does Ann have more than Pete?
Compare 3	Pete has 3 (5) apples; Ann has 6 (8) more apples than Pete; how many apples does Ann have?

* The numbers in brackets were used during the second and the third interview.

The word problems were read aloud by the interviewer. For each problem the children were asked to perform the following tasks: (1) to retell the problem, (2) to solve it, (3) to explain and justify their solution methods, (4) to build a material representation of the story with puppets and blocks, and (5) to write a matching number sentence. If the child did not succeed in solving the problem more or less independently, the interviewer switched over to the so-called systematic help procedure; this consisted in reading the problem sentence-by-sentence and asking the child after each sentence to represent the situation with the puppets and the blocks. The individual interviews were videotaped. The data were submitted to a quantitative and a qualitative analysis.

Two tasks in the individual interview were especially inserted to obtain information about the first stage of the problem-solving process in which the child constructs an internal representation of the problem, namely retelling the problem and building a material representation of it with puppets and blocks. The aim of the other tasks—explaining and justifying the solution strategy, and writing a matching number sentence—was to reveal children's thinking processes during the next stages of the problem-solving model, namely the selection and the execution of the formal arithmetical operation or the informal counting strategy.

Elsewhere we have given a systematic overview of the findings of this study with respect to children's problem representations (De Corte and Verschaffel 1985*a*, 1985*b*) as well as concerning their solution strategies (De Corte and Verschaffel 1984, 1985*b*); some practical implications of our work have also already been published (see De Corte and Verschaffel 1985*c*; De Corte, Verschaffel, Janssens, and Joillet 1985). The present paper focuses on a methodological question, namely whether retelling data yield useful and valid information about the internal problem representations that children construct during the first phase of the solution process of simple word problems. We will subsequently discuss the usefulness of retelling protocols as data, and some problems relating to the application of this technique. Beforehand, we will situate this question in a broader perspective, namely the value of verbal reports as data in research on children's thinking and problem-solving processes.

Retelling protocols and other verbal reports as data

During the last decade verbal reports, such as thinking aloud and retrospection protocols, have been applied more and more frequently in research on learning and instruction. This growing emphasis on protocol methods is related to an important shift in the object of instructional psychology. After a long era during which stimulus–response relations were at the focus of attention, psychologists are now trying to reveal the internal processes and cognitive structures underlying human behaviour. For that purpose it was necessary to apply and elaborate techniques that can provide data on those processes and structures, such as thinking aloud and retrospection protocols.

Although increasingly popular, protocol methods are not without their critics. The classical objections can be summarized as follows: (1) Because a subject may fail in verbalizing a considerable part of his thinking process, verbal reports provide an incomplete record of the thinking processes that are being or have been executed. (2) A subject's verbalizations may not coincide with his 'real' thought processes; therefore verbal reports may provide irrelevant and unreliable information. (3) Eliciting verbal reports may change a subject's 'normal' way of attacking and solving a problem, and hence provide an inaccurate picture of the course and the structure of the cognitive processes under study.

Until recently, there was almost no published literature on the methodological aspects of the collection and analysis of verbal reports, and, in research publications that made use of such data, the details of the methods were reported very sketchily (Ericsson and Simon 1980, p. 216). In view of the widespread and still increasing application of protocol methods, some efforts have been undertaken over the past few years towards a more thorough methodological analysis and justification of these techniques (Adair and Spinner 1981; Ericsson and Simon 1980, 1984; Ginsburg, Kossan, Schwartz, and Swanson 1983).

Undoubtedly, the most important contribution in this connection was made by Ericsson and Simon (1980, 1984). Based on the general framework of human information-processing theory on the one side, and on a thorough review of the available theoretical and empirical literature about the use of verbal protocols on the other, these authors proposed 'a model for the verbalization processes of subjects instructed to think aloud, to give retrospective reports, or to produce other kinds of verbalizations in response to experimenters' instructions' (1980, p. 217). This model provides a detailed description of the mechanisms and cognitive processes that occur when subjects are verbalizing or commenting on their ongoing or past cognitive activities. On the basis of this model one can predict when verbal reports will provide rich, reliable, and valid information and under which conditions this will not be the case. We will now summarize some main implications of Ericsson and Simon's (1980, 1984) model.

With respect to the completeness and reliability of protocol methods, the model assumes that only information that has been attended to in short-term memory can be reported accurately. Due to its limited capacity, information previously stored in short-term memory may be lost as new information is attended to; therefore 'only the most recently heeded information is accessable directly' (Ericsson and Simon 1980, p. 223).

An important distinction relates to the way in which information is encoded in short-term memory. When information is encoded verbally, the verbalization is a direct articulation or explication of the stored information; in that case, verbalizing can be done quite straightforwardly and easily. However, when information in short-term memory is not verbally encoded (e.g. visual imagery), verbal reporting requires intermediate recoding processes, which will make at least modest demands on processing capacity and processing time and may therefore cause inaccurate reports.

With respect to the issue of the influence of verbalizing instructions on subjects' performances on the task, Ericsson and Simon's (1980, 1984) analysis reveals that mere verbalizing does not significantly change the course and the structure of the thinking process, at least if the information is encoded in short-term memory either verbally or in a way that can be transformed easily into the verbal form.

The work of Ericsson and Simon clearly shows that a process model of verbalization is appropriate to analyse and evaluate thinking aloud, retrospection, and other procedures using verbal reports in the light of the criticisms that have been made of them.

When we apply Ericsson and Simon's model to the retelling data collected in our study of young children's internal representations of elementary arithmetic word problems, it seems that these data satisfy several conditions for obtaining reliable and valid data specified in the model. The children were asked to retell a word problem that was presented orally immediately before; it was assumed that in this situation they had to verbalize information encoded either verbally or in a way that can easily be transformed into the verbal mode. Further, as the retelling instruction was given immediately after the reading of the problem, it is plausible that this information was still available in short-term memory at the time of retelling.

On the other hand, the application of Ericsson and Simon's process model of verbalization to our retelling data is not without problems. First, the model itself still has some deficiencies and vaguenesses. For example, characteristics of the subject to whom the verbalization instructions are given (e.g. his active knowledge of vocabulary and grammatical rules, his metacognitive knowledge, his role-taking capacity, . . .) are almost totally undiscussed in their model. Second, our insufficient knowledge of the processes that occur when children try to understand and solve elementary arithmetic word problems makes it difficult to apply certain aspects of Ericsson and Simon's model to our verbalization procedure. For example, we do not yet have a clear picture of the content and the organization of children's short-term memory during the first stage of the solution process.

The usefulness of retelling protocols as data

We assumed that a qualitative analysis of the retelling protocols and of the material problem representations of the good as well as of the weak problem-solvers would teach us how they represented the problem situation internally. More specifically, we expected that those data would allow us to verify to what degree children's views of the problem situation correspond to the problem representations constructed by the computer simulation models of word-problem-solving processes developed by Riley *et al.* (1983) and by Briars and Larkin (1984).

In this section it will be shown that retelling data can be very helpful in generating and/or testing hypotheses about young children's representations of simple addition and subtraction word problems. We will subsequently discuss the following topics: (1) appropriate representations underlying correct solutions of elementary arithmetic word problems; (2) inappropriate representations underlying correct solutions; (3) inappropriate representations underlying incorrect answers on elementary arithmetic word problems.

Appropriate problem representations underlying correct solutions

First, we report some retelling data that allow us to conclude that children's appropriate representations of simple arithmetic word problems do not always coincide with the hypothetical representations implemented in the above-mentioned computer models. The following example relates to change-1 problems. Other illustrations are given in Verschaffel (1984).

In the computer models of Riley *et al.* (1983) and of Briars and Larkin (1984) it is assumed that for problems like our change-1 problem ('Pete had 3 apples; Ann gave Pete 5 more apples; how many apples does Pete have now?') a representation is constructed in terms of the change schema comprising three sets: a start set (3) and a change set (5), that are given, and a result set being the unknown quantity (see Fig. 3.1).

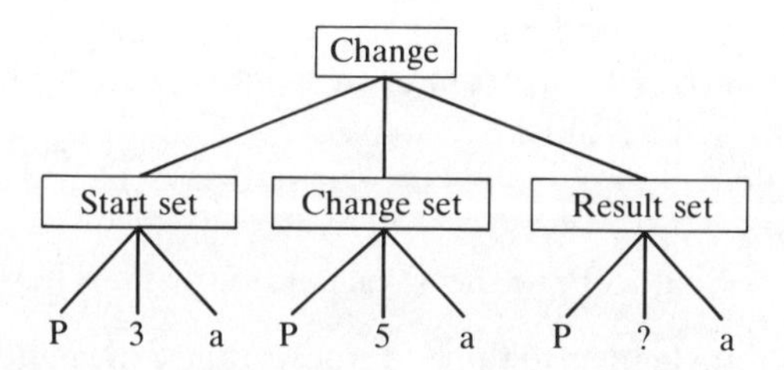

Fig. 3.1. Representation of the change-1 problem in terms of the elementary change schema (P: Pete; a: apple).

The retell protocols of many children who correctly solved the change-1 problem were in accordance with this hypothesis. Those retell protocols were very similar to the verbal text of the problem (e.g. 'Pete had 3 apples; he got 5 more apples; how many apples does Pete have now?'). In rendering the story with puppets and blocks these children carried out precisely the same material actions as described in Riley *et al.*'s (1983) and in Briars and Larkin's (1984) computer programs: first they took three blocks and put them with the puppet that represents Pete; then they added five more blocks to this set of three.

However, not all the good performers retold and materialized the

change-1 problem as described in the computer simulation models. In the retell protocol of ten pupils there was one more sentence than in the original problem text, in which those children stated that Ann initially had also available a specified or an unspecified number of apples. The following two examples were registered during the first session:
'Ann had 5 apples and Pete had 3 apples; Ann gave 5 apples to Pete; how many apples does Pete have now?'
'Pete had 3 apples and Ann had also some apples; she gave 5 of them to Pete; how many apples does Pete have now?'
Almost all children who retold the problem in such a way also played the story differently with the manipulatives. To start with they did not only put blocks with the puppet representing Pete, but also with Ann; then they moved five blocks from Ann towards Pete. In the comments accompanying their materializations those children oftentimes did not only refer to the apples that Ann had in the beginning, but also to those she still had available after giving five apples to Pete. The following protocol taken from the second session illustrates this obviously.

Protocol 1

Interviewer: Can you play the story with puppets and blocks?
Child: (Takes three blocks; puts them with the puppet which represents Pete, and says at the same time:) Pete had 3 apples.
(Then the child takes five blocks and puts them with Ann; next she moves them from Ann towards Pete while saying:) and Ann gave 5 blocks away.
(The child now looks to Ann and says:) and now Ann has no apples left.
(Looking to Pete she says:) and now Pete has 8 apples.

These empirical data lead us to the hypothesis that some children construct a more elaborated representation of change-1 problems than the elementary change schema described by Riley *et al.* (1983) and by Briars and Larkin (1984). Starting from the verbal text in which only three quantities are explicitly mentioned, they seem to construct a problem representation composed of five sets: (1) a set with the number of apples which Pete had available initially; (2) a set with the number of apples which Ann had at first; (3) a set with the number of apples transferred from Ann to Pete; (4) a set with the number of apples which Pete has at the end; (5) a set with Ann's final number of apples. The problem representation of our change-1 problem in terms of this more elaborated change schema is given in Fig. 3.2.

Similar data have been obtained for other change problems (see Verschaffel 1984). These findings justify the conclusion that retelling protocols provide useful, reliable, and valid data to reveal children's problem representations. On the one hand, they are helpful in discovering the variety in those representations, and, as such, they produce results that complement the findings obtained by means of other techniques such as computer simulation. On the other hand, the reliability and validity of the retell protocols is evidenced by the fact that problem representations emerging from those protocols are convergent with data collected by another technique, namely playing the story with puppets and blocks.

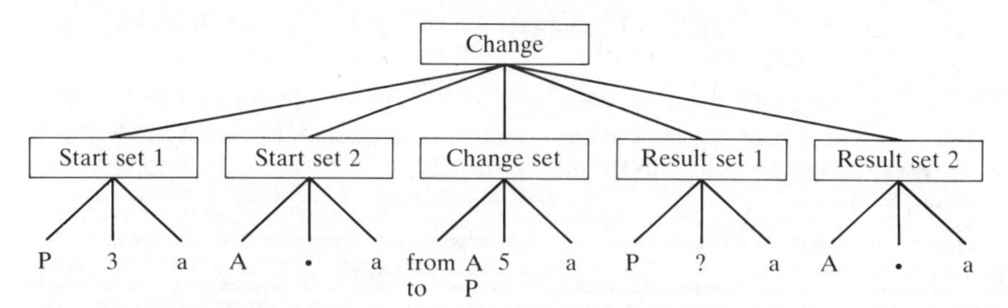

Fig. 3.2. Representation of the change-1 problem in terms of the elaborated change schema (P: Pete; A: Ann; a: apples).

Incorrect representations underlying correct solutions

In our competent problem-solving model the selection of the correct arithmetic operation to solve the problem is based upon an appropriate representation of the essential elements and relations in the problem situation. However, this does not imply that all correct solutions result from an appropriate problem representation. Indeed, a qualitative analysis of the retelling data of the good problem-solvers, together with their performances on the materialization task, clearly showed that some children answered a problem correctly without having a correct representation. Interestingly, almost all of these cases related to the difficult addition problems change-6 and compare-3 (Verschaffel 1984). We will illustrate this exemplarily for the latter problem type.

During the first, the second, and the third interview, respectively 5, 10, and 21 children correctly answered the compare-3 problem ('Pete has 3 apples; Ann has 6 more apples than Pete; how many apples does Ann have?') either independently or with a little help from the interviewer (Verschaffel 1984). However, during the second and the third interview, this correct answer was sometimes given immediately before or after the problem was retold in one of the following ways:

'Pete has 3 apples; Ann has 6 apples; how many apples do they have altogether?'

'Pete had 3 apples; Ann gave him 6 more apples; how many apples does Pete have now?'

These retelling protocols suggest that those children misinterpreted the difficult compare-3 problem in terms of the more familiar combine or change schema, respectively; nevertheless they obtained a correct answer because the value of the unknown set in their incorrect representation accidentally was the same as the good response on the compare-3 problem, namely 9.

This hypothesis about the origins of some children's correct solution on the compare-3 problem was confirmed by their performances on the materialization task. All children who answered the compare-3 problem correctly after retelling it wrongly in one of the above-mentioned ways failed to play the story with puppets and blocks. Therefore, the interviewer reread the problem sentence-by-sentence asking the child after each sentence to represent the situation with the manipulatives. After the first sentence ('Pete has 3 apples') they counted three blocks and put them with the puppet that represents Pete; after the second sentence ('Ann has 6 more apples than Pete') they counted a group of six blocks and put them with Ann; in reaction to the third sentence ('how many apples does Ann have?') they exchanged their original correct solution—9—for the faulty answer 6. For a detailed discussion of the mechanisms that can generate this wrong answer (6) we refer to another contribution (De Corte and Verschaffel 1985*a*). For the present discussion it is important that these data of the materialization task confirm the hypothesis based on the retelling protocols, namely that, in spite of their initial correct answer on the compare-3 problem, these children did not have an appropriate representation of the problem situation.

The preceding results support also the view that retelling protocols are useful in studying children's problem representations. More specifically, they can reveal whether a child's correct answer on a simple addition or subtraction word problem is based on a correct understanding of the problem situation. Again the convergence with the materialization data validates the retelling protocols.

Inappropriate representations underlying incorrect solutions

Analysis of children's performances on the retelling and materialization tasks can also be very helpful in discovering deficiencies in children's problem representations leading to logical and predictable errors. A number of the misconceptions that we found correspond to those described and modelled in the computer models of Riley *et al.* (1983) and Briars and Larkin (1984). However, some other deficiencies that we discovered are not implemented in these models. We will show this exemplarily for the combine-2 problems; similar data are presented elsewhere (De Corte and Verschaffel 1985*a*, 1985*b*).

According to Riley *et al.*'s (1983) and Briars and Larkin's (1984) computer models, children's failures on combine-2 problems are due to a lack of understanding of the part–whole relation. In traditional school-like combine problems this relationship is not stated explicitly in the verbal text. Consequently pupils who do not yet master the part–whole schema interpret each sentence of the word problem separately and cannot infer the relations between the two given sets. For example, when they are given the problem 'Tom and Joe have 8 marbles altogether; Joe has 5 marbles; how many marbles does Tom have?', they put out a set of eight blocks to represent the mables that Joe and Tom have altogether, and a separate set of five blocks to represent Joe's marbles. This leads to the incorrect response 8 when asked how many marbles Tom has.

During the first, the second, and the third interview, respectively nine, seven, and two children answered the combine-2 problem ('Pete has 3 apples; Ann has also some apples; Pete and Ann have 9 apples altogether; how many apples does Ann have?') as predicted by the computer models, namely with the largest given number, 9. However, an analysis of their retell protocols together with their materializations of the problem situation revealed that almost all these largest-given-number errors had an origin different from the one described in those models. We present here only one of the alternative origins, referring to another article for a more detailed discussion of the conceptual errors on combine-2 problems (see De Core and Verschaffel 1985*a*).

During the first and the second interview a few children answered the combine-2 problem with 9 after they retold it in one of the following ways*:
Pete had 3 apples; Ann had also some apples; Ann had 9 apples; Pete also had 9 apples; . . . (do you remember the question?). . . . How many apples did Ann have?'
'. . . . (Can you retell the problem?). . . . I only remember that Pete and Ann both had 9 apples.'
We hypothesize that these children failed on the combine-2 problem because they misinterpreted the sentence containing the word 'altogether' in the following way: 'Pete and Ann both possess 9 apples'. In other words, a statement containing the word 'altogether' is erroneously conceived as information concerning the amount of each person's property. The incorrect answer 9 is then the predictable consequence of this faulty interpretation of the third sentence in the problem. Again our explanation of this error based on the retelling data is nicely confirmed by those children's material representations: first they counted three blocks and put them with the puppet representing Pete; then they added blocks to this set until it contained nine blocks; thereafter they counted nine more blocks and put them with the puppet representing Ann.

* The words in brackets are said by the interviewer.

Problems related to the use of retelling protocols

The preceding section supports the view that retelling protocols can be valid and reliable data—that are very helpful in generating and/or verifying hypotheses about children's representations of simple arithmetic word problems. However, our analysis also shows that the use of this technique and the interpretation of the data it yields involve some specific problems. Therefore one cannot rely solely on a child's retelling protocol to derive the content and structure of its problem representation.

Incorrect retelling protocols in spite of appropriate problem representations

Several children failed on the retelling task, although they seemed to have an appropriate representation of the essential elements and relations in the problem.

Memory effects provide a plausible explanation for this finding. Retelling a word problem involves the reproduction of the sequence of sentences constituting the problem text; this reproduction is based on the representation the subject has built while listening to the verbal text. Especially for young children, this is a complex cognitive activity, putting a heavy load on working memory. Therefore, errors in a child's retelling protocol may occur that have nothing to do with misunderstanding the problem situation. The following protocol illustrates this obviously.

Protocol 2

Interviewer: The next problem is: Pete had some apples; he gave 5 apples to Ann; now Pete has 7 apples; how many apples did Pete have in the beginning? Can you retell this problem?

Child: Pete had some apples; he gave some apples to Ann . . . no . . . he gave 5 apples to Ann; how many apples did Pete have in the beginning?

Interviewer: Can you solve this problem already or do you want me to reread it?

Child: (Whispers:) He gave 5 apples away; then he still had 7 apples left; so he started with. . . .
(Says aloud:) 12.

In this child's retelling protocol one sentence is completely missing, namely the one in which the size of the result set is mentioned. The rest of the protocol shows, however, that this piece of information is nevertheless available and is, moreover, used appropriately during the problem-solving process. This justifies the conclusion that it was part of the problem representation. Probably the child just forgot to mention that part of his problem representation during the retelling task.

In the same way, insufficient mastery of lexical or syntactic rules or a deficient word-problem schema, which involves knowledge of the structure of arithmetic word problems, their intent, and their underlying assumptions (see the introductory section of this paper), may also cause errors in the retelling protocol of children who have constructed a correct representation of the problem situation (Verschaffel 1984).

Correct retelling protocols in spite of an inappropriate problem representation

During the analysis of our data we also found a few examples of children who correctly retold a word problem without having an appropriate internal representation of its essential elements and relations. The following protocol illustrates this obviously.

Protocol 3

Interviewer:	Listen to the next problem: Pete has 3 apples; Ann has 6 more apples than Pete; how many apples does Ann have? Can you retell this problem?
Child:	Pete has 3 apples; Ann has 6 apples. . . .
Interviewer:	I will read the problem once again: Pete has 3 apples; Ann has 6 more apples than Pete; how many apples does Ann have?
Child:	Pete has 3 apples; Ann has 6 more apples than Pete; how many apples does Ann have?
Interviewer:	Very well. Now try to solve it.
Child:	—
Interviewer:	Do you know the answer?
Child:	No.
Interviewer:	I will help you. I will read the problem now sentence-by-sentence; after each sentence you must try to represent the situation with the blocks. O.K.? Pete has 3 apples.
Child:	(Counts three blocks and puts them with the puppet representing Pete.)
Interviewer:	Ann has 6 more apples than Pete.
Child:	More. . . . I do not understand what that means.
Interviewer:	What do you not understand?
Child:	The sentence: Ann has 6 more apples than Pete.
Interviewer:	Can you put blocks with Ann so that she has 6 more than Pete?
Child:	(Counts six blocks and puts them with Ann but hesitates.)
Interviewer:	Does Ann have 6 more apples than Pete now?
Child:	I do not understand the meaning of that sentence.

We hypothesize the following explanation of this and similar cases. After hearing a text, a child has not only constructed a semantic representation, but, in addition, his working memory may also contain non-propositional information about graphical, lexical, and/or grammatical peculiarities of the text. To retell the text the child may rely on these more superficial memory traces (Kintsch, 1977), and this can result in a correct reproduction of the story in spite of an inappropriate semantic representation of the problem situation. The probability of such a correct reproduction increases as a function of the size of the text on the one hand, and the time interval between its presentation and the retelling instruction on the other. Our word problems were relatively short and the children were instructed to retell them immediately after they were read. Therefore we cannot preclude the possibility that some children succeeded in retelling a problem without having represented it appropriately.

Influence of the retelling instruction on children's problem-solving processes

One could argue that the retelling task can bear a significant influence on the manner in which children tackle word problems. The instruction to paraphrase the problem before solving it could alter the child's spontaneous listening and problem-solving strategies. More specifically, the retelling task might induce in children a more attentive and thoughtful approach than they usually apply. The so-called key-word strategy is a typical example of a superficial approach to word problems: the child's selection of the arithmetic operation is not based on a global analysis of the problem situation, but on the occurrence of isolated key-words in the verbal text (e.g. the words 'more' and 'less' are associated with addition and subtraction, respectively). When this strategy is applied to solve a problem in which the key-word does not refer to the correct arithmetic operation, it results in choosing a wrong operation (i.e. adding the two given numbers in the problem instead of subtracting, or the reverse); for example: 'Pete has 9 apples; he has 3 more apples than Tom; how many apples does Tom have? Answer: $9+3=12$ (Carpenter, Hiebert, and Moser 1983; De Corte, Verschaffel, and Verschueren 1982; Nesher and Teubal 1975). Although it seems reasonable to assume that the instruction to retell the word problem before solving it bears a significant influence on children's understanding and problem-solving processes, this hypothesis should be tested systematically in future research.

Lack of information concerning the process of problem representation

While the retelling task can be a very useful technique in reconstructing the product of a child's representation processes of a verbal problem, it does not yield information about those processes themselves. For example, using retelling protocols one cannot distinguish the relative importance of bottom–up and top–down analysis in the construction of an appropriate problem representation. Furthermore, an incorrect retelling protocol does not inform us about the origins of a child's misinterpretation, which can be very different; e.g. memory effects, the absence of the necessary schemata. . . .

Therefore, the application of a variety of data-gathering techniques is necessary to clarify the processes involved in the construction of children's problem representations. In accordance with this viewpoint we are now engaged in a study in which we collect eye-movement data while children read and solve word problems, together with their retrospections, in order to get a more complete picture of those processes.

Summary

In this paper we have discussed the usefulness and the validity of retelling data in studying young children's thinking processes with respect to elementary arithmetic word problems. This discussion was based on data collected during a longitudinal investigation in which thirty first-graders were given a series of eight simple addition and subtraction word problems three times during the school year. In order to get as rich and valid data as possible about all aspects of the thinking process, the children were given different tasks with respect to each word problem. The aim of the retelling task was to obtain information on how children represent such problems. Our analysis shows that retelling data provide rich and interesting information about children's problem representations. Moreover, when evidence from other data-gathering techniques was available, it mostly confirmed the retelling data. On the other hand, some findings clearly demonstrate at the same time the necessity of using simultaneously other methods of data collection. Indeed, it is obvious that one cannot rely on retelling as the only source of data on children's cognitive processes during problem-solving. However, taking into account the unquestionable value of retelling protocols as data, the problems related to collecting and interpreting them deserve further methodological research.

References

Adair, J. G. and Spinner, B. (1981). Subjects' access to cognitive processes: demand characteristics and verbal reports. *Journal for Theory of Social Behavior* **11,** 31–52.

Briars, D. J. and Larkin, J. H. (1984). An integrated model of skill in solving elementary word problems. *Cognition and Instruction* **1,** 246–96.

Carpenter, T. P., Hiebert, J., and Moser, J. M. (1983). The effect of instruction on children's solutions of addition and subtraction word problems. *Journal for Research in Mathematics Education* **14,** 55–72.

De Corte, E. and Verschaffel, L. (1984). First graders' solution strategies of addition and subtraction word problems. In *Proceedings of the Sixth Annual Meeting of the North American Chapter of the International Group for the Psychology of Mathematics Education* (ed. J. M. Moser), pp. 15–20. Wisconsin Center for Education Research, University of Wisconsin, Madison, WI.

De Corte, E. and Verschaffel, L. (1985*a*). Beginning first graders' initial representation of arithmetic word problems. *Journal of Mathematical Behavior* **4,** 3–21.

De Corte, E. and Verschaffel, L. (1985*b*, March). *An empirical validation of computer models of children's word problem solving.* Paper presented at the Annual Meeting of the American Educational Research Association, Chicago, IL.

De Corte, E. and Verschaffel, L. (1985*c*). Working with simple word problems in early mathematics instruction. In *Proceedings of the Ninth International Conference for the Psychology of Mathematics Education. Vol. 1. Individual Contributions* (ed. L. Streefland), pp. 304–9. Research Group on Mathematics Education and Educational Computer Center, Subfaculty of Mathematics, State University of Utrecht, Utrecht, The Netherlands.

De Corte, E., Verschaffel, L., and Verschueren, J. (1982). First graders' solution processes in elementary word problems. In *Proceedings of the Sixth International Conference for the Psychology of Mathematical Education* (ed. A. Vermandel), pp. 91–6. Universitaire Instelling Antwerpen, Antwerpen, Belgium.

De Corte, E., Verschaffel, L., Janssens, V., and Joillet, L. (1985). Teaching word problems in the first grade: a confrontation of educational practice with the results of recent research. In *Using research in the professional life of mathematics teachers* (ed. T. A. Romberg), pp. 186–95. Wisconsin Center for Education Research, University of Wisconsin, Madison, WI.

Ericsson, K. A. and Simon, H. A. (1980). Verbal reports as data. *Psychological Review* **87,** 215–51.

Ericsson, K. A. and Simon, H. A. (1984). *Protocol analysis. Verbal reports as data.* MIT Press, Cambridge, MA.

Ginsburg, H. P., Kossan, N. E., Schwartz, R., and Swanson, D. (1983). Protocol methods in research on mathematical thinking. In *The development of mathematical thinking* (ed. H. P. Ginsburg), pp. 7–47. Academic Press, New York.

Greeno, J. G. (1978, March). *Significant basic research questions and significant applied research questions are the same questions.* Paper presented at the Annual Meeting of the American Educational Research Association, Toronto, Ontario, Canada.

Heller, J. I. and Greeno, J. G. (1978, May). *Semantic processing of arithmetic word problem solving.* Paper presented at the Annual Meeting of the Midwestern Psychological Association, Chicago, IL.

Kintsch, W. (1977). On comprehending stories. In *Cognitive processes in comprehension* (eds. M. A. Just and P. A. Carpenter), pp. 33–62. Erlbaum, Hillsdale, NJ.

Nesher, P. and Teubal, E. (1975). Verbal cues as an interfering factor in verbal problem solving. *Educational Studies in Mathematics* **6,** 41–51.

Riley, M. S. and Greeno, J. G. (1978, May). *Importance of semantic structure in the difficulty of arithmetic word problems.* Paper presented at the Annual Meeting of the Midwestern Psychological Association, Chicago, IL.

Riley, M. S., Greeno, J. G., and Heller, J. I. (1983). Development of children's problem-solving ability in arithmetic. In *The development of mathematical thinking* (ed. H. P. Ginsburg), pp. 153–96. Academic Press, New York.

Verschaffel, L. (1984). *Representatie- en oplossingsprocessen van eersteklassers bij aanvankelijke redactie-opgaven over optellen en aftrekken. Een theoretische en methodologische bijdrage op basis van een longitudinale, kwalitatief-psychologische studie (First graders' representations and solution processes on elementary addition and subtraction word problems. A theoretical and methodological contribution based on a longitudinal, qualitative-psychological investigation).* Unpublished Doctoral Dissertation. Seminarie voor Pedagogische Psychologie, Faculteit der Psychologie en Pedagogische Wetenschappen, K. U. Leuven, Belgium.

4

Understanding of arithmetical operations as models of situations

BRIAN GREER

Introduction

In the 1980s there has been a considerable body of research into children's understanding of how addition, subtraction, multiplication, and division model various situations. This contrasts with the bulk of earlier research on the operations which dealt with such topics as the learning and retention of the number facts, and performance of computational algorithms. Such research accurately reflected the predominant assumption that teaching arithmetic is primarily a matter of teaching computational skills. This assumption is now being challenged by many people working in mathematics education who believe that the operations should be taught, within a wide range of contexts, as models of situations.

Work on two specific areas within the general topic of the application of the four operations to the real numbers is reviewed here. The choice of these areas—addition and subtraction of small positive integers, modelling situations involving numbers of discrete objects; and multiplication and division of positive integers and decimals, modelling a variety of situations—reflects concentrations of recent research. While there are considerable differences between the two areas, notably in the ages of pupils studied and in methodologies, several common psychological and pedagogical themes emerge in the three sections which follow, dealing in turn with findings, theories, and implications.

Findings

Problem taxonomies and comparisons of difficulty

Addition and subtraction

Numerous researchers have shown that the difficulty of problems modelled by the same calculation can vary enormously. Vergnaud (1982, p. 38) gives the following illustration of three problems all solved by the addition 4 + 7 but differing markedly in difficulty:

Problem A: There are 4 boys and 7 girls round the table. How many children are there altogether?

Problem B: John has just spent 4 francs. He now has 7 francs in his pocket. How much did he have before?

Problem C: Robert played two games of marbles. On the first game, he lost 4 marbles. He played the second game. Altogether, he now has won 7 marbles. What happened in the second game?

Out of such research on the psychologically distinguishable situations modelled by addition and subtraction have evolved various (largely overlapping) classification schemes. As Vergnaud (1982) remarks, such schemes are helpful in interpreting what pupils do, in understanding their difficulties, in providing a framework for thinking about the domain, and in designing systematic experiments. Vergnaud provides a classification and a schematic representation; classification schemes are also detailed in Carpenter and Moser (1983), Nesher, Greeno, and Riley (1982), and Riley, Greeno, and Heller (1983). Table 4.1 summarizes some of the main distinctions that have been made. Change problems are those which can be conceptualized in terms of a transformation of some initial quantity. Combine problems are those which can be conceptualized in terms of joining disjoint sets or partitioning a set into two subsets. Compare problems are those which can be conceptualized in terms of a relationship between two quantities. (Not included in Table 4.1 is a further type, equalizing—see Carpenter and Moser (1983)).

In Table 4.1, problems are also distinguished notationally as follows:

STS: an initial state (*S*) is subject to a transformation (*T*) which results in another state;

SSS: two states are combined, resulting in a third state;

SRS: a relation (*R*) exists between two states.

The unknown quantity in each case is enclosed in square brackets. It will be seen that change and compare problems are grouped in threes according to which quantity is the unknown, but that combine problems are different in that the symmetry of roles played by the two subsets means that two of the cases are equivalent (as indicated in Table 4.1).

Children from kindergarten to grade 3 have been studied. In most of the research, the method used has been to read problems to children who are then asked to solve them. Often concrete objects have been available to use as counters; De Corte and Verschaffel (in press) took this further by providing puppets as concrete referents for people mentioned in the problems. In addition to observation of overt behaviour and recording of verbal responses, probing questions and retrospective reports on solution processes have been used.

Various consistent differences have been found between problems. Representative results from a study by Riley (cited in Riley *et al.* 1983) have been

Table 4.1. *Types of addition and subtraction problems*[a].

	ST[S][b]	*S[T]S*	*[S]TS*
Change (add)	Joe had 5 marbles. Ann gave him 3 more marbles. How many marbles does Joe have now? 87/100/100/100[c]	Joe had 3 marbles. Then Ann gave him some more marbles. Now Joe has 8 marbles. How many marbles did Ann give him? 61/56/100/100	Joe had some marbles. Then Ann gave him 3 marbles. Now Joe has 8 marbles. How many marbles did Joe have at first? 9/28/80/95
Change (subtract)	Joe had 8 marbles. Then he gave 3 marbles to Ann. How many marbles does Joe have now? 100/100/100/100	Joe had 8 marbles. Then he gave some marbles to Ann. Now Joe has 3 marbles. How many marbles did he give to Ann? 91/78/100/100	Joe had some marbles. Then he gave 3 marbles to Ann. Now Joe has 5 marbles. How many marbles did Joe have at first? 22/39/70/80
	SS[S]	*S[S]S*	
Combine	Joe has 5 marbles. Ann has 3 marbles. How many marbles do they have altogether? 100/100/100/100	Joe and Ann have 8 marbles altogether. Joe (Ann) has 3 marbles. How many marbles does Ann (Joe) have? 22/39/70/100	
	SR[S]	*S[R]S*	*[S]RS*
Compare (greater)	Joe has 5 marbles. Ann has 3 more marbles than Joe. How many marbles does Ann have? 13/17/80/100	Joe has 8 marbles. Ann has 3 marbles. How many marbles does Joe have more than Ann? 17/28/85/100	Joe has 8 marbles. He has 3 more marbles than Ann. How many marbles does Ann have? 17/11/65/75
Compare (less)	Joe has 8 marbles. Ann has 3 fewer marbles than Joe. How many marbles does Ann have? 17/28/90/95	Joe has 3 marbles. Ann has 8 marbles. How many marbles does Joe have less than Ann? 4/22/75/100	Joe has 5 marbles. He has 3 less marbles than Ann. How many marbles does Ann have? 0/6/35/75

[a] Problems with 5 and 3 are addition, problems with 8 and 3 subtraction.
[b] See text for explanation of notation.
[c] Figures are percentages correct for Kindergarten and for grades 1, 2, and 3, respectively (data of Riley cited in Riley *et al.* 1983).

included in Table 4.1. A particularly reliable finding is that problems in the first column, where the final state is the unknown, are easier than the corresponding problems in the other columns. Within addition problems in the first column, compare problems have been found to be more difficult than change and combine problems. Within subtraction problems, combine and most of the compare problems have been found to be more difficult than the change (subtract) problem in the first column. Summaries of other findings will be found in Carpenter and Moser (1983) and Riley *et al.* (1983).

The situations considered here by no means exhaust the possibilities. Vergnaud (1982) exemplifies a number of others. For instance, his problem C, quoted at the start of this section, may be characterized as a *T[T]T* problem, that is to say, the combination of two transformations (the second unknown) into a single transformation. Moreover, consideration has been limited to situations involving sets of discrete objects. Sooner or later, addition and subtraction have to be understood also in relation to ordinal structures and directed numbers (Bell 1985) and measures. Freudenthal (1983) raises the question of problems such as: 'John is 5 years old. How old will he be in 3 years time?' and asks (p. 98): 'How is the arithmetical knowledge about addition that was acquired by uniting sets transferred didactically to this kind of problem?'

Multiplication and division

Whereas addition and subtraction take place within a single measure-space, multiplication and division involve up to three distinct measure-spaces (e.g. distance, time, and speed); moreover, various different relationships between the measure-spaces are possible. A preliminary attempt at a classification of some of the main categories of problem is shown in Tables 4.2 and 4.3. Table 4.2 includes cases for which multiplication is 'psychologically non-commutative'. Take the first problem: '3 boys had 4 marbles each. How many marbles did they have altogether?' In such problems the number of groups and the number of objects in each group play quite different roles. The situation is naturally conceived of as '3 lots of 4'; thus 4 is the *multiplicand,* and 3 is the *multiplier* which operates on it to produce the answer. As a consequence, two types of division are distinguishable, as indicated in Table 4.2. Division by the number of groups is called *partition,* while division by the number in each group is called *quotition.*

In each of the multiplication situations listed in Table 4.2, one of the quantities can be identified as the multiplicand and the other as the multiplier. In Table 4.2, therefore, the concepts of partitive and quotitive division have been extended by defining partition generally as division by the multiplier and quotition as division by the multiplicand (Greer and Mangan 1984).

In the categories in Table 4.3, by contrast, the roles played by the

Table 4.2. *Types of multiplication and division problems (asymmetrical cases).*

Category	Multiplication	Division (partition)	Division (quotition)
Multiple groups	3 boys had 4 marbles each. How many marbles did they have altogether?	12 marbles were divided equally among 3 boys. How many marbles did they get each?	12 marbles were divided among some boys. Each boy got 4 marbles. How many boys were there?
Iteration of measure	4 pieces of wood are each 3.2 m long. What is the total length of wood?	A piece of wood 12.8 m long is cut into 4 equal pieces. How long is each piece?	A piece of wood 12.8 m long is cut into pieces 3.2 m long. How many pieces are obtained?
Change of scale	In a photograph, the length of a car is 3.2 cm. If the photograph is enlarged by a factor of 4.5, how long will the car be in the enlarged photograph?	A photograph is enlarged by a factor of 4.5. In the enlarged photograph a car is 14.4 cm long. How long is the car in the original photograph?	In a photograph, a car is 3.2 cm long. The photograph is enlarged and in the enlarged photograph the car is 14.4 cm long. What is the enlargement factor?
Rate	A man walks for 4.5 hours at a steady speed of 3.2 m.p.h. How far does he walk?	A man walks 14.4 miles in 4.5 hours. What is his speed in m.p.h?	A man walks 14.4 miles at a steady speed of 3.2 m.p.h. How long does it take him?
Measure conversion	If the rate of exchange is 1.5 dollars per pound, how many dollars will you get for £3.20?	If you get 4.80 dollars for £3.20, what is the exchange rate in dollars per pound?	If the rate of exchange is 1.5 dollars per pound, how many pounds will you get for $4.80?

quantities multiplied are symmetrical, as indicated by the examples. This distinction between the two sets of situations is similar to that between change and compare problems, on the one hand, and combine problems on the other, for addition and subtraction (Table 4.1).

In terms of the notation used previously, the problems in Table 4.2 may be seen as analogous to *STS* problems in addition and subtraction; the multiplication problems correspond to type *ST[S]*, partitive division to type *[S]TS*, and quotitive division to type *S[T]S* (in relation to this see discussion of Vergnaud's (1983) theory below). Problem types in Table 4.3 are roughly

Table 4.3. *Types of multiplication and division problems (symmetrical cases).*

Category	Multiplication	Division
Rectangular array	If there are 3 rows and 4 columns, what is the total number?	If the total is 12 and there are 3 rows (columns), how many columns (rows) are there?
Combinations	If there is a choice of 3 colours and 4 styles, how many combinations of colour and style are there?	If there are 12 combinations of colour and style and there are 3 choices of colour (style), how many choices of style (colour) are there?
Area	If the length is 3.2 cm and the breadth is 4.5 cm, what is the area?	If the area is 14.4 cm^2 and the length (breadth) is 3.2 cm what is the breadth (length)?

analogous to the *SSS* type, but less closely because of the added dimensional complexity. Other types not in the tables include *TTT* problems, which could arise, for example, for any of the situations in Table 4.2 in which the result is a measure on the same dimension as the multiplicand—thus, two changes of scale can be combined into a single change of scale.

On the basis of more formal criteria, Vergnaud (1983) distinguished three classes of multiplicative structure, as follows:

(1) Isomorphism of measures, involving direct proportion between two measure-spaces, M_1 and M_2. This category subsumes all the types listed in table 4.2. For example, for a problem involving constant speed, M_1 is the time and M_2 the distance travelled.

(2) Product of measures, namely the composition of two measure-spaces, M_1 and M_2, into a third, M_3. This category subsumes all the types listed in Table 4.3.

(3) Multiple proportion, in which a measure-space M_3 is independently proportional to two other measure-spaces M_1 and M_2. This is exemplified by a measure such as man-hours.

By contrast with the work on addition and subtraction, most of the research on multiplication and division has been done with children in the 12–14 age-range. The experimental approaches used in this research also differ somewhat. Most of the emphasis has been on whether or not the child can choose the correct operation to solve a given problem. Carrying out the computation has usually not been required—the child simply has to state which computation would give the answer. In pencil-and-paper tests the children are asked to write down the calculation (Bell, Fischbein, and Greer

1984; Fischbein, Deri, Nello, and Marino 1985) or to choose from a fixed set of alternatives (Ekenstam and Greger, 1983; Hart, 1981). Clinical interviews have also been extensively used (Bell *et al.* 1984; Bell, Swan, and Taylor 1981; Ekenstam and Greger 1983; Greer in press). A second major difference is that a lot of attention has been given to the effects on problem difficulty of the types of number used.

Some illustrative results from paper-and-pencil choice of operation tests are presented in Tables 4.4, 4.5, and 4.6. Table 4.4 shows percentages for various multiplication problems used in an experiment with thirty 12- and 13-year-olds of slightly above-average ability (Bell *et al.* 1984). Judging from these examples, multiple groups and iteration of measures present no difficulty. The first two rate problems suggest that when the multiplication involves one decimal (greater than one) and one integer it makes a big difference which is the multiplicand and which the multiplier. The final rate problem and the change-of-scale problem indicate difficulty when the multiplier is a decimal less than one.

Table 4.4. *Choice of operation: multiplication (examples from Bell et al. 1984).*

Category	Problem	% correct
Multiple groups	Tins of beans are packed in cardboard boxes. Each box contains 24 tins. If a boy unpacks 16 boxes, how many tins of beans has he unpacked?	100
Iteration of measure	In the school kitchen the cooks use 0.62 kg of flour to make one tray of doughnuts. How much flour will it take to make 27 trays?	97
Change of scale	To fit a picture of a dress onto a page of a magazine, the picture has to be reduced to 0.14 of its original size. In the original picture the length of the dress was 2 m. What will its length be in the magazine?	10
Rate	My car can go 41.8 miles on each gallon of petrol on a motorway. How many miles can I expect to travel on 8 gallons?	94
Rate	An international cross-country runner completed a training run in 1.13 hours. He maintained an average speed of 9 m.p.h. How long was the course?	63
Rate	One gallon of petrol costs £1.33. How much will it cost to fill up a small tank which can only hold 0.53 gallons?	27

Table 4.5. *Choice of operation: division (examples from Bell et al. 1984).*

Category*	Problem	% correct
Multiple groups (Q)	I am packing boxes of cassettes. I fit 24 cassettes in a box. I have 600 cassettes. How many boxes will I fill?	87
Iteration of measure (P)	25 men do the football pools together. This week they have won £22. How much will they receive each?	70
Rate (Q)	One pound of meat is priced at £2.56 in a shop window. A housewife pays £2.00 for a cut of meat. What should it weigh?	23
Measure conversion (Q)	A table is 92.3 cm long. About how many inches is this? (1 inch is about 2.54 cm).	70
Measure conversion (Q)	The petrol tank of a Mini holds 5.5 gallons. How much is this in litres? (One litre is 0.22 gallons).	43

* Q represents quotition, P partition.

Table 4.5 shows some division problems from the same study. Again, the multiple-groups problem proved easy. On the second question, 27 per cent of the pupils gave the answer as 25/22 and, on the third, 30 per cent gave the answer as 2.56/2, suggesting that division of a smaller number by a larger may be a source of difficulty leading to the order of the numbers being reversed. The final two questions illustrate the added difficulty of dividing by a decimal less than one as opposed to one greater than one.

The above findings have to be treated cautiously since the questions differed in many respects, making any inferences based on comparisons tentative. Table 4.6 shows some results from a study in which more-or-less matched pairs of problems were given to fifty-six 13- and 14-year-olds of below-average ability (Bell 1984). The first two pairs show clearly that partition is much more difficult when the divisor is larger than the dividend. The other pairs show that multiplication and division are more difficult when the operator is a decimal less than one as opposed to a decimal greater than one.

Findings from a more systematically designed study have been briefly reported by Greer and Mangan (1984). A total of 144 12- and 13-year-olds of about-average ability were tested on closely matched problems of the types change of scale, measure conversion, and two types of rate problem—time/distance/speed and unit price/amount/total cost. No real differences in difficulty were found between these problem types, but very clear results emerged relating to the effects on difficulty of the types of number involved

Table 4.6. *Choice of operation: specific comparisons (examples from Bell 1984).*

Category*	Problem	% correct
Iteration of measure (P)	If 13 churns of milk hold 164 gallons, how much milk is in each churn?	63
	If 40 cups of tea are poured from an urn holding 25 pints, how much tea is in each cup?	23
Iteration of measure (P)	If 13 fishing-boats are to tie up alongside a quay 164 m long, what length of quay is available for each fishing-boat?	66
	If 40 model ships placed end-to-end measure 25 m altogether, how long is each one?	23
Rate	I buy some coffee beans costing £0.88 per lb. If I buy 1.6 lb, how much will I have to pay?	39
	A woman buys some meat costing £0.88 per lb. If she buys 0.8 lb, how much does it cost?	20
Rate (Q)	A car can travel 37.5 miles on one gallon of petrol. How much petrol does it use on a journey of 56 miles?	32
	A car can go 37.5 miles for every gallon of petrol. How much petrol does it use on a journey of 28 miles?	14
Change of scale	A photograph of the Eiffel Tower is to be used for a poster. Every length in the poster is 1.25 as long as the corresponding part in the photograph. If the tower measures 13 cm from top to bottom in the photograph, what will it measure on the poster?	18
	To fit a picture of a dress onto a page of a magazine, the picture has to be reduced to 0.75 of its original size. In the original picture the dress was 14 cm long. What will its length be in the magigine?	5

* Q represents quotition, P partition.

in the problems. For multiplication, three levels of difficulty were found depending on the type of number used as multiplier. When it was an integer, 92 per cent of the answers were correct; when it was a decimal greater than one, 71 per cent; and when it was a decimal less than one, 53 per cent. By contrast, the type of number used as multiplicand had very little effect. The theoretical implications of this are considered later.

Division problems were much more difficult (33 per cent correct answers vs. 75 per cent for multiplication). Again, difficulty was very largely dependent on the types of numbers involved. Partitive and quotitive problems were of about the same level of difficulty.

In several studies a different slant has been provided by reversing the task demand, i.e. by giving the children calculations and asking them to make up word problems which fit them (Bell *et al.* 1984; Ekenstam and Greger 1983; Greer and Mangan 1984; Hart 1981). Roughly speaking, what these studies show is that if the calculations are such that they can be fitted by problems of the multiple-groups or iteration-of-measure type, appropriate problems of such types are given with a reasonable degree of success. Other calculations are generally handled badly; often they are 'forced' into forms easier to deal with; e.g. by using a decimal number in a role which logically demands an integer.

Semantic and linguistic factors in solving word problems

Apart from the analysis of semantic categories which has already been discussed, there are many specific semantic and linguistic factors implicated in the solution of verbally presented problems. Nesher (1982) pointed out that addition and subtraction word problems have a well-defined minimal structure consisting of three underlying strings, two providing information and one specifying the unknown information to be derived. The logical relations between these can be linguistically encoded in a variety of manners, of which Nesher lists seven: arguments, adjective, agents, location, time, verbs, relational terms. The first of these refers to 'semantic dependence among the numerically quantified arguments occurring in the strings underlying the problem text', as in this example (p. 31): 'Three *boys* and two *girls* went to the beach. How many *children* went to the beach?' 'Verbs' refers to the fact that semantic dependence may be implied by verbs in the text, as in the example (p. 31): 'Victor *had* five stamps and *gave* two of them to Joe. How many stamps does Victor *have* now?' Another important distinction is between dynamic, static, and comparison texts (closely related to the change, combine, and compare semantic categories).

Other studies have looked at more surface features of texts. The most obvious of these is the existence of 'key-words' as cues for particular operations. Children are often taught to look for such words—'altogether' in the question indicates that addition (or multiplication) is required, etc. Yet it is not difficult to find examples where this strategy breaks down. Consider these four problems used by Nesher and Teubal (1975, p. 51):

A. The milkman brought on Sunday 4 bottles of milk more than on Monday. On Monday he brought 7 bottles. How many bottles did he bring on Sunday?

B. The milkman brought on Monday 7 bottles of milk. That was 4 bottles less than he brought on Sunday. How many bottles did he bring on Sunday?

C. The milkman brought 11 bottles of milk on Sunday. That was 4 more than he brought on Monday. How many bottles did he bring on Monday?
D. The milkman brought on Sunday 11 bottles of milk and on Monday he brought 4 bottles less. How many bottles did he bring on Monday?

In A, the key word 'more' suggests addition, which is in fact the correct operation. However, 'more' also occurs in C, which requires subtraction. Likewise, 'less' appears in D (solved by subtraction) and in B (solved by addition). Results of tests showed that correct answers were much more frequent for A and D, where the key-words are consistent with the operation, than for B and C, where they are not.

Nesher and Teubal (1975) distinguish three levels in the transition from natural language to corresponding symbolic expressions: (a) the verbal formulation; (b) the underlying mathematical relations; (c) the symbolic mathematical expression. As they point out, the key-word strategy is based on the fallacy that a direct correspondence can be established between (a) and (c).

The four problems listed above illustrate another aspect, namely 'the temporal sequentiality of the events, and the flow of information in the verbal text' (Nesher 1982, p. 36). In A, C, and D, the description of what happened on Sunday precedes in the text the description of what happened on Monday. Thus the order in which the information is presented is consistent with the order of events described. In problem B, however, the reverse is the case. De Corte and Verschaffel (in press) also draw attention to this aspect as worthy of more investigation.

Some of the most interesting findings, with profound educational implications which are considered later, are those showing a close relationship between semantic categories of problems and children's methods of constructing the answers. Hiebert (1984, p. 501) gives the following example:

1. Tom has 13 pennies. He gives 8 pennies to Jane. How many pennies does Tom have left?
2. Tom has 8 pennies. Jane gave him some more and now he has 13 pennies. How many pennies did Jane give to Tom?

Most first-graders provided with objects as counters will solve the first by removing 8 counters from a set of 13, but the second by adding counters to a set of 8 until there are 13. As Hiebert says, this shows that 'they are able to analyze a word problem and extract its intended meaning, rather than searching for a key word or carrying out some other superficial analysis' (p. 501).

A relationship between semantic categories and solution methods was found by Carpenter and Moser (1982) for subtraction problems but not for addition problems. Using more sensitive methodology, De Corte and Verschaffel (in press) found it for both subtraction and addition problems. They also extended the scope of the finding by providing evidence that the relationship exists not only for solution strategies inferred from overt behaviour and verbal responses, but also for mental solutions, as inferred from retrospective verbal reports. A particularly interesting illustration of this is the following:

1. Pete has 5 apples; Ann has 9 apples; how many apples do Pete and Ann have altogether?
2. Pete had 5 apples; Ann gave Pete 7 more apples; how many apples does Pete have now?

Problem 2 is a change problem and changing the order of numbers would conflict with the semantic structure. Most children who solved problem 2 mentally reported that they counted on from 5, the *first* number. By contrast, problem 1 is a combine problem; since the two numbers play symmetrical roles, there is no psychological obstacle to changing the order of the two numbers to use the more efficient strategy of counting on from the *larger* number, and most children who solved the problem mentally reported doing this. Thus the children's mental solution processes were sensitive to the semantic structure of the problem. (This example also serves to illustrate the distinction made between symmetrical and asymmetrical situations in Table 4.1).

Although much less work has been done on the effects of semantic and linguistic factors in relation to multiplication and division problems, it is unlikely that they are any less important in that context.

Theories

History of reinforcement: a simple explanation?

In explaining the experimental findings that have been summarized here, quite a lot of mileage can be got out of a straightforward explanation in terms of the children's experience in school. For example, in the Concepts in Secondary Mathematics and Science Project (reported in Hart 1981), misconceptions about multiplication and division were identified, namely that 'multiplication makes bigger, division makes smaller' and that division is always of the larger number by the smaller. While more precise delineation is necessary of the circumstances in which these misconceptions come into play, they have been observed in many studies. The point relevant to the present argument is that these misconceptions, which account for many of the errors on multiplication and division word problems, can be explained as incorrect generalizations from extensive early experience of calculations confined to positive integers, within which domain they are correct.

With respect to conceptual understanding of situations modelled by the operations, it is clear that most children's experience is limited and highly biased. De Corte, Verschaffel, Janssens, and Joillet (1984) surveyed instructional programmes used with Belgian first-grade classes and found an imbalance in the proportions of addition and subtraction problems from different semantic categories, with a marked preponderance of change and combine problems with the result unknown. They concluded that this would tend to produce superficial solution strategies for the types overrepresented and inability to cope with those underrepresented or not represented at all.

As well as being extremely imbalanced, word problems are generally stereotyped (Nesher 1980). The result of this is that pupils evolve 'short cuts' to identify the operation required on the basis of 'surface structure' rather than by understanding the 'pragmatic deep structure'. Sowder, Threadgill-Sowder, Moyer, and Moyer (in press) have identified a number of strategies used by children in deciding which operation to use:

A. Find the numbers and add.
B. Guess the operation.
C. Calculate all the possibilities and choose the most reasonable.
D. Decide the size of the answer relative to the operand. If larger, try addition and multiplication and choose the more reasonable; if smaller, try subtraction and division and choose the more reasonable.
E. Look for a 'key-word' signalling the correct operation.

A further strategy observed (Sowder, personal communication) is to infer the operation from the numbers; e.g. if they are 78 and 54, it is probably addition or subtraction, whereas if they are 78 and 3, it is probably division.

A major part of the problem is that these strategies can be undeservedly successful. Schoenfeld (1982) reports that in a widely used elementary textbook series, 97 per cent of the problems could be 'correctly solved' by the key-word strategy.

Another aspect of increasing importance is the availability of computer software. Microcomputers offer scope for presenting a wider range of situations, with the extra potential offered of using dynamic displays to represent spatial and temporal relationships. On the other hand, they could merely provide a more efficient way of reinforcing the existing imbalance. In the mathematical section of '*The which? software guide*' (Walker 1984), three programs are described which represent arithmetical operations graphically (along with 17 which are designed to practice calculation). One represents addition by showing a crane put blocks onto two trucks, subtraction by blocks being lifted off a boat, and addition and subtraction by ducks being added to or taken off a river. Another represents addition by apples falling from two sides of a tree. All of these fall within the categories identified by De Corte *et al.* (1984) as dominating first-grade instruction (see above). A third program represents division by an octopus sharing out suitably marine objects, an example of arguably the most primitive conception of division.

Primitive intuitive models: Fischbein's theory

Fischbein *et al.* (1985) have put forward a theory to account for findings from studies of multiplication and division word problems. Rather than attribute these simply to the children's experience in school, as considered above, Fischbein *et al.* postulate a more fundamental basis for the weaknesses in conception and performance. The hypothesis is stated thus:

> 'Each fundamental operation of arithmetic generally remains linked to an implicit, unconscious, and primitive intuitive model. Identification of the operation needed to solve a problem with two items of numerical data takes place not directly but as mediated by the model. The model imposes its own constraints on the search process' (p. 4).

The primitive model for multiplication is assumed to be repeated addition. In the simplest situation (multiple groups in Table 4.2), '3 lots of 4' is seen as $4 + 4 + 4$. This can easily be extended to iteration of measures; e.g. '3 lots of 4.2 m' is $4.2 + 4.2 + 4.2$ m. It can be further extended to the other situations in Table 4.2; e.g. a man walking for 3 hours at 4.2 m.p.h. walks $4.2 + 4.2 + 4.2$ miles. The crucial factor is that the *multiplier* must be an integer; the type of number used as the *multiplicand* is relatively unimportant. Experimental evidence consistent with this has been reported above. From the viewpoint of the theory, the 'multiplication-makes-bigger' misconception is a logical by-product of multiplication being conceived of as repeated addition. The fact that multipliers which are decimals greater than one are coped with much better than those which are less than one is explained by the notion that the decimal parts of such numbers are 'absorbed' into the integral parts.

For division, there are two primitive models: partition (equal sharing) and quotition (how many x's in y?). Within partitive division, according to the theory: (a) the dividend must be larger that the divisor, (b) the divisor must be an integer, (c) the result must be smaller than the dividend (which can account for the 'division-makes-smaller' misconception). (The necessity of the first of these is not obvious, since the notion of sharing can easily be extended to cases such as four cakes divided among six children.)

For quotitive division the only constraint assumed is that the dividend must be larger than the divisor (which can account for the 'larger-divided-by-smaller' misconception).

There is considerable evidence from choice-of-operation tests, making-up-story tests, and interviews which is consistent with this theory (Bell *et al.* 1984; Greer and Mangan 1984). However, most, if not all, of the same predictions can be made on the basis of the experiential factors considered in the previous section. It may be questioned whether the primitive models of the operations postulated are the only ones possible. For example, the possibility of a conception of multiplication as enlargement is not considered. However, the general point seems incontestable, that the conceptualization of the operations is fundamentally affected by the situations embodying them which are encountered first, and most commonly (both in and out of school).

Some specific aspects of the hypothesis are debatable—for example, the statement that repeated addition is the model 'that tacitly affects the meaning and use of multiplication, even in persons with considerable training in mathematics' (Fischbein *et al.* 1985)—this remains to be empirically tested. Another is the exclusive emphasis on enactive representations, which ignores the possibility of ikonic representations, such as the rectangular array for multiplication (see Table 4.3) and movements on the number line for addition and subtraction (Freudenthal 1983).

The theory raises the question as to what extent the effects that have been described are the result of teaching and to what extent a natural cognitive development. Fischbein *et al.* (1985) tend to the pessimistic view that the primitive models are inevitable and that the teacher's job is 'to attempt to provide learners with efficient mental strategies to control the impact of these primitive models' (p. 16). A more positive strategy would be to aim to widen the range of models available to the pupils.

A broader perspective: Vergnaud's analysis

Vergnaud's three-way classification of multiplicative structures was outlined earlier. Here we will concentrate on his analysis of the first category, isomorphism of measures. For situations involving a direct proportion between two measure-spaces, M_1 and M_2, Vergnaud sees multiplication and division problems as special cases of the general class of 'rule-of-three' problems for which one of the terms is equal to one. In these terms, multiplication, partitive division, and quotitive division problems are as shown in Fig. 4.1 (a)–(c). In each case, there are potentially two routes to the solution, as indicated. For example, the result of a multiplication can be derived by noting that the transformation which takes 1 to e within M_1, namely multiplication by e, must be applied within M_2 to d. Alternatively, the transformation which takes 1 to d, namely multiplication by d, may be applied to e. Findings from various experiments by Vergnaud (and others) support his contention that children have a very marked preference for thinking in terms of transformations *within* measures rather than transformations *between* measures. In Vergnaud's terms, multiplication by the scalar operator, e, is much more natural than multiplication by the function operator, d. This provides an alternative explanation for the findings reported earlier concerning the psychological non-commutativity of multiplication. In fact, what Vergnaud terms the function and scalar operators correspond to what were earlier characterized as the multiplicand and multiplier, respectively.

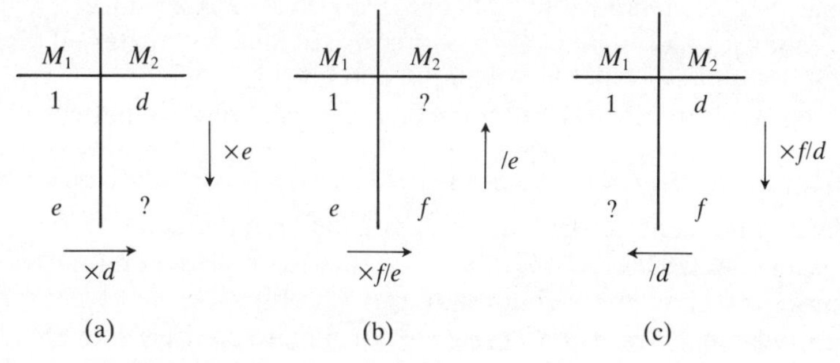

Fig. 4.1. Isomorphism of measures: (a) multiplication, (b) partitive division, (c) quotitive division.

Fig. 4.1 (b) and (c) make plain that there is a structural difference between partitive and quotitive divisions. In terms of transformations within measure-spaces, quotitive division involves multiplication by f/d (applied to 1), whereas partition involves division by e (applied to f). However, it is not clear what predictions can be made on the basis of this analysis for differences in performance on problems involving the two types of division.

The foregoing analysis supports the previous characterization of situations in Table 4.2 as *STS* problems, the two measures within M_2 being the states, and the scalar operator the transformation.

(It should be noted that this summary covers only part of Vergnaud's (1983) wide-ranging analysis of multiplicative structures, which is placed within a much broader context than has been attempted in this paper. From his standpoint, multiplication and division form part of a 'conceptual field' which includes many other concepts such as fraction, ratio, similarity, etc. Addition and subtraction, likewise, form part of a complex conceptual field.)

Computer models

Two sets of precise models of how addition and subtraction word problems are solved have been developed as computer simulations (Briars and Larkin 1984; Riley *et al.* 1983). The models developed by Riley *et al.* are constructed in terms of a general model of problem-solving based on (a) problem schemata—for representing knowledge about the semantic relations in the problem, (b) action schemata—for representing knowledge about actions involved in solution processes, and (c) strategic knowledge for planning solutions. A problem is solved by first constructing a representation of its semantic structure, then generating a solution guided by the planning procedures.

Models within this general framework have been developed for change, combine, and compare problems. Three levels of performance are postulated. At level 1, the child's representations are limited to external displays of concrete objects available as counters; at level 2, schemata are available to represent additional information about the relationships between quantities; at level 3, there are additional schemata enabling the representations and solution processes to be carried out in a more top–down fashion; i.e. by establishing the general structure first and filling in the details later.

In particular, at level 3 there is a schema for representing part–whole relationships which allows children to solve problems of the types:
Joe had some marbles. Then Tom gave him 5 more marbles. Now Joe has 8 marbles. How many marbles did Joe have in the beginning?
Joe had some marbles. Then he gave 5 marbles to Tom. Now Joe has 3 marbles. How many marbles did Joe have in the beginning?
For the former, at level 2, there is no capability for recording the unknown start set, and the change set of five objects is represented incorrectly as the start set and given as the answer. At level 3, the model infers that the start and change sets are subsets of the result set and further infers from this that the start set must consist of the extra objects added to the change set to make a total of eight. In terms of actions with objects, the solution can be constructed either by putting out five objects, then counting how many more are needed to get to eight, or by putting out eight and removing five of them.

The model of Briars and Larkin (1984) hypothesizes the same three levels of performance. The general cognitive processes underlying the performances are also similar, except at level 3. Here Briars and Larkin postulate a subset-equivalence schema (somewhat similar to Riley's part–whole schema) which allows children to interchange subsets. This makes it possible for children to use the more efficient counting-on-from-larger procedure when the second number is the larger (see the example from De Corte and Verschaffel (in press) cited in the section on semantic factors (p. 71); also Baroody and Ginsburg (in press)). It also allows solution of change (add) problems with the initial quantity unknown (the first type exemplified above) but not of change (subtract) problems with the initial quantity unknown (the second type exemplified above). For those, an extra time-reversal schema is proposed which allows transformations to be reversed in time (cf. Bell 1985).

How much empirical support is there for these models? There is quite a lot of evidence that patterns of performance over problems of different semantic structures are similar for the models, and for children. Of course, much of this evidence is simply the evidence on the basis of which the details of the models were derived in the first place. Findings inconsistent with the models have been reported by De Corte and Verschaffel (this volume). Carpenter and Moser (1984) report similar misgivings and question whether such models in general are able to cope with the variability in children's performance; the same point is made by Baroody and Ginsburg (in press). Moreover, the fact that Riley *et al.* (1983) and Briars and Larkin (1984) make the same predictions for level 3 performance based on different assumptions about the underlying cognitive processes is a reminder that these models provide sufficient, but not necessary, explanations for the performance they produce.

Educational implications

Several coherent themes emerge from the two bodies of research reviewed here despite the differences between them in methodology and in the level of complexity of the problems. These may be summarized as follows:

1. Sets of word problems which in a sense are of the same type vary a great deal in difficulty. For addition and subtraction problems the emphasis has been on how differences in semantic structure produce differences in difficulty of problems solved by the same calculation. For multiplication and division the emphasis has been on how differences in the numbers involved produce differences in difficulty in problems otherwise the same (see especially Greer (in press)).
2. The range of semantic structures children consistently interpret correctly is limited. This may be attributable to the excessive concentration in their education on the more obvious structures, to the virtual exclusion of others, or may have a more fundamental basis, as suggested by Fischbein *et al.* (1985).
3. Solution processes are rarely concept-driven (to use Sowder's term). Instead, they are based on superficial cues such as the types of numbers involved, or key-words. Such strategies are reinforced by success rates which can be very high.

Hiebert (1984) suggests that the root of the problem is the breakdown of communication between knowledge about form, and understanding. As one illustration of this thesis he cites the work on addition and subtraction which shows that children initially have considerable understanding of the mathematical relationships within situations, and considerable prowess in the use of informal solution methods, but that the effects of schooling are that 'many children mechanically add, subtract, multiply or divide whatever numbers are given in a problem with little regard for the problem's content . . . somehow in learning formal arithmetic procedures, many children stop analyzing the problems they attempt to solve' (Carpenter and Moser 1982, p. 9). These conclusions are totally in accordance with Donaldson's (1978) critique of premature formalization in school education in general.

What is to be done? There is widespread pressure for two fundamental changes in teaching approaches. The first is that the operations should be taught in relation to a wide variety of situations they model. Carpenter and Moser (1982, p. 9) argue that 'verbal problems may give meaning to addition and subtraction and in this way could represent a viable alternative for developing addition and subtraction concepts in school'. This is echoed by De Corte and Verschaffel (in press); moreover, they have implemented a first-grade programme based on these principles, which shows evidence of substantial improvement in children's performance (De Corte and Verschaffel 1985).

The second, related, change that is urged is that children's informal solution methods should be respected and the introduction of formal methods (the ultimate need for which is not disputed) should be carried out much more gradually and sensitively.

Many other aspects could be mentioned:

(a) Teaching materials (including computer software) need to be monitored to check for balanced sampling of problem types and types of number combinations.

(b) Teachers need to become adept at diagnosing, exposing, and eliminating misconceptions.

(c) Specific attention needs to be paid to the critical periods during which the domain of an operation is expanded—notably when addition and subtraction are extended from positive to directed numbers, and when multiplication and division are extended from integers to decimals.

(d) Much more research is needed on visually mediated processes in solving problems, particularly the use of schematic diagrams and the potential of dynamic computer graphics.

Acknowledgement

My work in this area is largely the result of collaboration with colleagues at the Shell Centre for Mathematical Education in Nottingham, in particular Alan Bell.

References

Baroody, A. J. and Ginsburg, H. P. (in press). The relationship between initial meaningful and mechanical knowledge of arithmetic. In *Conceptual and procedural knowledge: the case of mathematics* (ed. J. Hiebert). Erlbaum, Hillsdale, NJ.

Bell, A. W. (1984). Structures, contexts and learning: some points of contact between cognitive psychology and mathematical education. *Journal of Structural Learning* **8,** 165–71.

Bell, A. W. (1985, March). *Understanding additive situations involving directionality–relations between context, structure, and language.* Paper presented to a symposium on Cognitive Processes in Mathematics and Mathematics Learning, University of Keele, Keele, England.

Bell, A. W., Fischbein, E., and Greer, G. B. (1984). Choice of operation in verbal arithmetic problems: the effects of number size, problem structure and context. *Educational Studies in Mathematics* **15,** 129–47.

Bell, A. W., Swan, M., and Taylor, G. (1981). Choice of operation in verbal problems with decimal numbers. *Educational Studies in Mathematics* **12,** 399–420.

Briars, D. J. and Larkin, J. H. (1984). An integrated model of skill in solving elementary word problems. *Cognition and Instruction* **1,** 245–96.

Carpenter, T. P. and Moser, J. M. (1982). The development of addition and subtraction problem-solving skills. In *Addition and subtraction: a cognitive perspective* (eds. T. P. Carpenter, J. M. Moser, and T. A. Romberg). Erlbaum, Hillsdale, NJ.

Carpenter, T. P. and Moser, J. M. (1983). The acquisition of addition and subtraction problems. In *Acquisition of mathematical concepts and processes* (eds. R. Lesh and M. Landau). Academic Press, New York.

Carpenter, T. P. and Moser, J. M. (1984). The acquisition of addition and subtraction concepts in grades one through three. *Journal for Research in Mathematics Education* **15,** 179–202.

De Corte, E. and Verschaffel, L. (1985). Working with simple word problems in early mathematics instruction. In *Proceedings of the Ninth International Conference for the Psychology of Mathematics Education. Vol. 1. Individual Contributions* (ed. L. Streefland), pp. 304–9. Research Group on Mathematics Education and Educational Computer Center, Subfaculty of Mathematics, State University of Utrecht, Utrecht, The Netherlands.

De Corte, E. and Verschaffel, L. (in press). First graders' solution strategies of elementary arithmetic word problems. *Journal for Research in Mathematics Education.*

De Corte, E., Verschaffel, L., Janssens, V., and Joillet, L. (1984). *Teaching word problems in the first grade: a confrontation of educational practice with results of recent research.* Paper presented at the Fifth International Congress on Mathematical Education, Adelaide, Australia.

Donaldson, M. (1978). *Children's minds.* Fontana, London.

Ekenstam, A. and Greger, K. (1983). Some aspects of children's ability to solve mathematical problems. *Educational Studies in Mathematics* **14,** 369–84.

Fischbein, E., Deri, M., Nello, M. S., and Marino, M. S. (1985). The role of implicit models in solving decimal problems in multiplication and division. *Journal for Research in Mathematics Education* **16,** 3–17.

Freudenthal, H. (1983). *Didactical phenomenology of mathematical structures.* Reidel, Dordrecht.

Greer, G. B. (in press). Non-conservation of mathematical operations. *Journal for Research in Mathematics Education.*

Greer, G. B. and Mangan, C. (1984). Understanding multiplication and division. In *Proceedings of the Sixth Annual Meeting of the North American Chapter of the International Group for the Psychology of Mathematics Education* (ed. J. M. Moser). Wisconsin Center for Education Research, University of Wisconsin, Madison, WI.

Hart, K. (ed.) (1981). *Children's understanding of mathematics: 11–16.* Murray, London.

Hiebert, J. (1984). Children's mathematics learning—the struggle to link form and understanding. *Elementary School Journal* **84,** 496–513.

Nesher, P. (1980). The stereotyped nature of school word problems. *For the Learning of Mathematics* **1,** 41–8.

Nesher, P. (1982). Levels of description in the analysis of addition and subtraction word problems. In *Addition and subtraction: a cognitive perspective* (eds. T. P. Carpenter, J. M. Moser, and T. A. Romberg). Erlbaum, Hillsdale, NJ.

Nesher, P. and Teubal, E. (1975). Verbal cues as an interfering factor in verbal problem solving. *Educational Studies in Mathematics* **6,** 41–51.

Nesher, P., Greeno, J. G., and Riley, M. S. (1982). The development of semantic categories for addition and subtraction. *Educational Studies in Mathematics* **13,** 373–94.

Riley, M. S., Greeno, J. G., and Heller, J. I. (1983). Development of children's problem-solving ability in arithmetic. In *The development of mathematical thinking* (ed. H. P. Ginsburg), pp. 153–96. Academic Press, New York.

Schoenfeld, A. H. (1982). Some thoughts on problem-solving research and mathematics education. In *Mathematical problem solving: issues in research* (eds. F. K. Lester and J. Garofalo). Franklin Institute Press, Philadelphia.

Sowder, L. K., Threadgill-Sowder, J. A., Moyer, J. C., and Moyer, M. B. (in press). Diagnosing a student's understanding of operations. *The Arithmetic Teacher.*

Vergnaud, G. (1982). A classification of cognitive tasks and operations of thought involved in addition and subtraction problems. In *Addition and subtraction: a cognitive perspective* (eds. T. P. Carpenter, J. M. Moser, and T. A. Romberg). Erlbaum, Hillsdale, NJ.

Vergnaud, G. (1983). Multiplicative structures. In *Acquisition of mathematics concepts and processes* (eds. R. Lesh and M. Landau). Academic Press, New York.

Walker, J. (1984). *The which? software guide.* Hodder and Stoughton, London.

5

Strategy choices in subtraction

ROBERT S. SIEGLER

Not only is it important to know more about each individual child's developmental rate in regards to mathematics, but also we should know more about each individual child's particular processes, number systems, 'easy' and 'hard' numbers and number combinations, short cuts, and devices which he uses in arriving at answers to arithmetic problems (Ilg and Ames 1951, p. 26).

Ilg and Ames' comment, which they made 35 years ago, reflects understanding of a point that has been ignored too often by more contemporary cognitive psychologists. Most cognitive psychological models depict performance in terms of a single strategy or procedure that people are said to use whenever they perform a task. However, careful observation of performance, such as that done by Ilg and Ames, reveals considerable variability in people's strategies. Even a single child performing a single problem twice within a single experiment sometimes uses two different strategies on the two occasions. For example, a boy trying to subtract 5 from 9 might retrieve an answer from memory and state it on an initial trial, yet when presented the same problem later in the experiment might count on his fingers. Such cases are not rare. For example, in the first experiment that is discussed below, children were presented with each subtraction problem twice within the experiment; they used different strategies on 34 per cent of the pairs of trials.

I recently have examined children's strategy choices in three arithmetic domains: addition, multiplication, and subtraction. In each domain, I have focused on four questions. What strategies do children use to solve problems of this type? What advantages do children gain from using diverse strategies? How do they choose which strategy to use on a given occasion? What implications do the strategy choice procedures have for learning? This paper is organized around these four questions as they apply to subtraction; for discussions of strategy choices in addition and multiplication, see Siegler and Shrager (1984) and Siegler (in press).

What strategies do children use to subtract?

Several previous investigators have noted that children use a variety of strategies to subtract. Starkey and Gelman (1982) noted that 3- to 5-year-olds put up fingers and count them to solve some subtraction items, count aloud without any obvious external referent on other subtraction items, and retrieve from memory the answers to yet other items. Ilg and Ames (1951), like Starkey and Gelman, observed that 5-year-olds sometimes use the counting fingers strategy and other times count in the same way but without putting up fingers. They also noted that 6- and 7-year-olds use these strategies and also retrieve some subtraction facts from memory. Woods, Resnick, and Groen (1975) inferred from reaction-time patterns that 7- and 9-year-olds sometimes count down from the larger number the number of times indicated by the smaller, and other times count up from the smaller number until they reach the larger. When counting down, the remaining portion of the larger number would be the answer; when counting up, the number of counts would be the answer. Svenson and Hedenborg (1979) inferred from reaction-time patterns that 9-, 10-, and 11-year-olds use the two strategies described by Woods *et al.* (1925) and also that they retrieve from memory the answers to several types of problems: problems where the larger and smaller numbers are equal, problems where the smaller number is 0, problems where the larger number is exactly twice as large as the smaller one, problems where the larger number is 10, and occasionally other problems as well. Lankford (1974) obtained verbal protocols from 13-year-olds in the process of solving subtraction problems and found that they too use both counting and retrieval strategies.

These studies clearly establish that at least from age 3 to age 13, children use a variety of strategies on simple subtraction problems. They leave unclear, however, the degree to which individual children use multiple strategies and the consequences of their using the multiple strategies. One way to establish these characteristics was to videotape children's subtraction performance. This was done in the present study.

Method

The children were seventeen 5-year-olds and seventeen 6-year-olds. The 5-year-olds attended a university pre-school. The 6-year-olds attended the first grade of an upper-middle-class suburban school. The numbers of boys and girls were almost identical within each age-group. The experimenter was a 27-year-old female research assistant.

Children were presented with the 25 subtraction problems in which the smaller number (the subtrahend) ranged from 1 to 5 inclusive, and the difference from 1 to 5 inclusive. Each child was presented the 25 problems over a 3-day period, with 8 or 9 problems per day; then the child was presented the same problems again, under the same circumstances but in a different order. Before the first session, children were told:

I want you to imagine that you have a pile of oranges. Then imagine that I take some oranges away from your pile. Tell me how many you have left. You can do anything you want to help you get the right answer. If you want to use your fingers or count aloud, that's fine. OK? Suppose you have *m* oranges, how many would you have if I took away *n* of them?

After phrasing the problems this way three times, the experimenter abbreviated the presentation to 'Suppose you had *m* and I took away *n*?'

Solution times were recorded through use of a digitizer that fed into the videocassette recorder and printed digital times across the bottom of the taped scene. The times were accurate to 1/10th of a second, which seemed a sufficient degree of accuracy for the present task where the average trial lasted for 6 seconds.

Results

Existence of strategies

Analysis of the videotapes indicated that the kindergarteners and first-graders used four strategies: counting fingers, fingers, counting, and retrieval. These were the same four strategies that Siegler and Shrager (1984) had observed in 4- and 5-year-olds' addition. The *counting fingers strategy* involved putting up fingers to represent the larger number (sometimes all at once, other times one at a time), putting down fingers until the number of fingers put down equalled the smaller number (usually done one at a time, occasionally done all at once), and finally counting the number of fingers in the difference. The *counting strategy* involved similar behaviours except that the children did not use fingers or other obvious external referents; they simply counted aloud. The *fingers strategy* involved putting up fingers corresponding to the larger number, putting down fingers corresponding to the smaller number, and then simply stating the answer without counting the number of fingers in the difference. Finally, the fourth strategy, *retrieval*, involved no overt counting and no use of fingers or other external objects; children simply stated the answer following presentation of the problem. Two independent raters agreed on 98 per cent of their classifications of the

Table 5.1. *Characteristics of subtraction strategies.*

Strategy	Trials on which strategy used (%)	Mean solution time (sec)	Correct Answers (%)
Counting fingers	21	9.8	83
Fingers	14	4.4	94
Counting	7	9.1	25
Retrieval	58	4.4	68

strategies. As shown in Table 5.1, retrieval was the most frequently used strategy, followed by counting fingers, fingers, and counting.

Individual children's patterns of strategy use

Individual children tended to use several different strategies, rather than just one. Only 12 per cent of the children (4 of 34) used a single strategy on the 50 trials; all these children used retrieval on all problems. Of the remaining children, 12 per cent used 2 different strategies at least once, 33 per cent used 3 different strategies, and 42 per cent used all 4 of the strategies. The most common combinations of strategies used by individual children were (a) retrieval, fingers, and counting fingers (27 per cent) and (b) all 4 strategies (42 per cent).

The above summary does not imply that children who used multiple strategies used the strategies equally often. Instead, almost all of them tended to use one or two strategies primarily. Children used whichever strategy they used most often on an average of 75 per cent of trials, and whichever two strategies they used most often on an average of 93 per cent of trials. For 69 per cent of children, the most frequently used strategy was retrieval, for 21 per cent it was counting fingers, for 6 per cent it was counting, and for 6 per cent it was fingers. Even among the 42 per cent of children who used all 4 strategies at least once, the most commonly used strategy accounted for 68 per cent of responses, and the two most common for 89 per cent. Thus, most children used three or four subtraction strategies at least sometimes, but most of them also relied predominantly on one or two strategies.

Relative solution times of strategies

The distribution of associations model, which will be described below, predicts that retrieval should require the shortest solution times, fingers the next shortest, and counting fingers and counting the longest. These predictions were tested by performing t-tests in which solution times of pairs of strategies were compared for those children who used each of the pair of strategies at least three times; that is, to be included in the analysis of the relative solution times of the retrieval and the fingers strategy, a child needed to use retrieval at least three times and to use the fingers strategy at least three times.

These analyses indicated that children executed the retrieval strategy significantly more quickly than the fingers strategy, $t(17) = 2.70$, that they executed the fingers strategy significantly more rapidly than the counting fingers or counting strategy, $t(17) = 5.77$, and that they did not differ significantly in the time they spent executing the counting fingers and counting strategies, $t(6) = 0.58$. These relations were exactly the ones found to exist among these four strategies in addition (Siegler and Shrager 1984).

The results of the experiment also allowed examination of the second main question that we have asked about strategy choices.

How does it help children to use multiple strategies?

The benefits of using multiple strategies emerge most clearly when both speed and accuracy are considered. An ideal single strategy for solving a set of problems would be both the fastest and the most accurate of all the strategies that could be used to solve the problems. However, no single subtraction strategy possesses both of these virtues. As noted above, children answered most rapidly when they used the retrieval strategy. Retrieval was not the most accurate strategy, though; both the fingers and the counting fingers strategies were associated with more accurate responding (Table 4.1).

Although the fastest strategy, retrieval, is not associated with the most accurate responding overall, it does lead to accurate responding on the easiest subtraction problems. For example, in an experiment described below in which 5- and 6-year-olds were required to use retrieval on all trials, they answered correctly on 86 per cent of the trials where they used the retrieval strategy to solve $4-1$, 85 per cent where they used it to solve $3-1$, and 84 per cent where they used it to solve $2-1$. They were only slightly more accurate on these problems in the present experiment when they used one of the three overt strategies (87 per cent, 88 per cent, and 88 per cent, respectively). Thus, unless accuracy was of the utmost importance, it would seem to be to their advantage to use retrieval most of the time on these problems, since they could be quick as well as accurate using it.

Conversely, it would seem to be the 5- and 6-year-olds' advantage to use *overt strategies* (strategies involving visible or audible behaviour, in this case counting, counting fingers, and fingers) most often on the problems where retrieval leads to relatively poor performance: problems such as $7-5$, $8-5$, and $9-5$. Children were substantially more accurate on these difficult problems when they used overt strategies than when they used retrieval (70 per cent vs. 50 per cent, 76 per cent vs. 50 per cent, and 75 per cent vs. 50 per cent, respectively).

In all of the experiments that my colleagues and I have conducted on strategy choices, children have shown an uncanny ability to use overt strategies most often on the problems where the overt strategies are most helpful (Siegler in press; Siegler and Robinson 1982; Siegler and Shrager 1984; Siegler and Taraban 1986). The wisdom of their strategy choices can be seen in the correlations between the frequency of their use of overt strategies on a given problem and the difficulty of the problem, measured either by the frequency of errors on the problem or by the solution time required to solve the problem. The more difficult the problem, the more often they use overt strategies. In the present study of subtraction, the 5- and 6-year-olds' frequency of overt strategy use on each subtraction problem correlated $r = 0.84$ with their percentage of errors on the problem and $r = 0.90$ with their mean solution times on it (Fig. 5.1). These strong relations are comparable to those reported by Siegler and Shrager (1984) for addition ($r = 0.91$ and $r = 0.92$, respectively) and by Siegler (in press) for multiplication ($r = 0.83$ and $r = 0.86$, respectively).

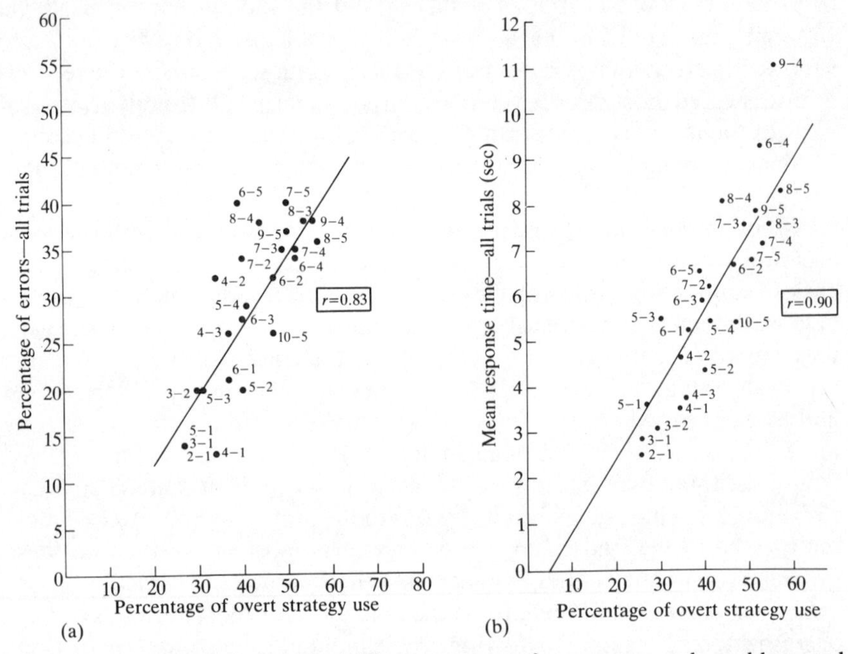

Fig. 5.1. Correlations between (a) percentage of errors on each problem and percentage of overt strategy use on that problem and (b) between mean solution times on each problem and percentage of overt strategy use on the problem.

One possibility was that these strong correlations were due to use of the overt strategies causing high frequencies of errors and long solution times. Examination of the errors and solution times on retrieval trials suggested that this was not the case, though. Performance on these retrieval trials was of special interest because, by definition, use of overt strategies could not have influenced errors and solution times on them. The correlation between the percentage of errors on retrieval trials on each problem and the percentage of overt strategy use on that problem was identical to that found when the percentage of errors on all trials for the problem was correlated with percentage of overt strategy use on the problem, $r = 0.83$. The correlation between mean solution times on retrieval trials on each problem and percentage of overt strategy use on each problem also remained substantial, again

$r = 0.83$. Thus, use of overt strategies did not appear to cause the strong correlations among errors, solution times, and overt strategy use. Instead, it appeared that overt strategy use, errors, and solution times all reflected the difficulty of the problem. The question then became how children knew to use overt strategies predominantly on the most difficult problems.

How children choose strategies: the distribution of associations model

The distribution of associations model was designed to explain not only how children know to use overt strategies most often on the most difficult problems, but also to explain a variety of other aspects of their performance: which strategies they use, the relative solution times of the strategies, the relative numbers of errors and length of solution times on different problems, and the close relations among errors, solution times, and overt strategy use on each problem. Before discussing how this model works, though, it seems worth-while to describe two previous models of young children's simple subtraction that can be contrasted with the current one.

Prior models of simple subtraction

Woods *et al.* (1975) proposed that elementary school-age children use two distinct procedures for subtracting. Which procedure they use is totally determined by the relation between the larger and the smaller numbers. Whenever the larger number is more than twice as great as the smaller (e.g. $8 - 2$), children count down from the larger number by one's the number of times indicated by the smaller number, reducing the smaller number by one at each count. Once the smaller number is exhausted, they state the remaining part of the larger number as the answer; that is, on $8 - 2$, 8 is greater than twice the smaller number, so children count down from 8 for 2 counts, thereby exhausting the smaller number, and state the remaining part of the larger number, 6, as the answer. The other procedure is used whenever the larger number is less than twice as great as the smaller. Here children count up by one's from the smaller to the larger number, keeping track of the number of counts. When the running total equals the larger number, the number of counts is the answer. For example, on $8 - 6$, 8 is less than twice the smaller number, so children count up from 6 for 2 counts. At this point, they state the number of counts, 2, as the answer.

The second model of single-digit subtraction, proposed by Svenson and Hedenborg (1979) is an elaboration of the Woods *et al.* (1975) model with a series of specific tests grafted onto the beginning of it. Children first test whether the larger and smaller numbers are equal; if so, they say the answer is 0. Next, they test whether the smaller number is 0; if so, they say the answer equals the value of the larger number. Next they check whether the larger number is exactly twice as great as the smaller; if so, they say that the answer equals the smaller number. Following this, they try to retrieve the answer from long-term memory. Next, they check whether the larger number is 10; if it is, they solve the problem quickly, either through retrieval or through an unspecified computational process. If they have not yet answered, they determine whether the larger number is greater than 10 and the smaller number is less than 10; if so, they always count down from the larger number the number of times indicated by the smaller. Finally, if none of these procedures yields an answer, they use the Woods *et al.* procedure to generate a solution.

Both of these models were designed to account for relative solution times on small-number subtraction problems. The Woods *et al.* model accounted for 56 per cent of the variance for those second-graders where it was the best-fitting model, and for 41 per cent of the variance in solution times for all of the fourth-graders who were tested. Svenson and Hedenborg's (1979) model accounted for 84 per cent of the variance in solution times for third-through fifth-graders. Before concluding that the Svenson and Hedenborg model is more successful in accounting for the data, however, note that this model included 7 fitted parameters, whereas the Woods *et al.* model included only 2.

Although these models account reasonably well for the children's solution times, they aslo have several limitations. One involves the limited range of data that they explain; they only account for relative solution times on different problems. A second, related problem concerns the difficulty of inferring solely from solution-time data children's mental processes. Consider the assumption of both models that children count up from the smaller number when the larger number is less than twice as great as the smaller, and that they count down from the larger number when the larger number is more than twice as great as the smaller. Children may in fact do this, but the solution times alone are inherently ambiguous evidence. The identical pattern of solution times would be produced if the children always counted down when subtracting, but kept track both of whether they had exhausted the smaller number and of whether they had reached the smaller number. If they exhausted the smaller number first (as on $8-2$), they would answer with the remaining part of the larger number; if they reached the smaller number first (as they would on $8-6$), they would answer with the number of counts downward they had made. This approach, in addition to being consistent with the data, would avoid the need to decide in advance whether $M > 2N$. Children may always use one approach, always use the other, or sometimes use one and sometimes the other; solution-time patterns alone cannot discriminate, though.

An additional difficulty of these models is their awkwardness in integrating retrieval and counting processes. Woods *et al.* (1975) noted that the fit of their model to second-graders' performance improved substantially when problems where the minuend equaled twice the subtrahend (8 – 4) were excluded. These problems, like the corresponding tie problems in addition, were solved much more quickly than the model predicted. Woods *et al.* suggested that answers to these problems were probably being retrieved, and also suggested that the decrease in solution times from second to fourth grade was probably attributable to increasing percentages of retrievals being mixed in with the counting-up and counting-down processes. Svenson and Hedenborg (1979) assumed that retrieval was always used on some problems, such as those where the minuend equalled twice the subtrahend, and that it was sometimes used on other problems as well. However, the solution times alone provided no basis for identifying just when retrievals occurred on these other problems. As Svenson and Hedenborg noted, if children retrieved answers to some of these problems more often than others (a possibility that seems more than likely), the effect would be to bias other parameter estimates within the model.

The distribution of associations model

The distribution of associations model was first proposed by Siegler and Shrager (1984) in the context of simple addition. It was designed to show how children's adaptive choices of strategies grow out of their basic procedures for solving arithmetic problems. It also was designed to show how learning of arithmetic problems is produced by the same processes as performance on them.

The distribution of associations model for subtraction is much like that proposed by Siegler and Shrager for addition. In both cases, the model is organized into a representation and a process. The representation consists of associations of varying strengths between each problem and possible answers to the problem. The numerical values in the Fig. 5.2 matrix are the estimated strengths of these associations. These estimated strengths are based on children's performance in a separate experiment, the overt-strategies-prohibited experiment. In this subtraction experiment, 94 5- and 6-year-old kindergarteners and the first-graders were presented the same 25 subtraction problems as in the above-described overt-strategies-allowed experiment, but were explicitly asked to 'just say what you think the right answer is without putting up your fingers or counting'. The purpose of these instructions was to obtain the purest possible estimate of the strengths of associations between problems and answers. The numbers in the matrix, the associative strengths, correspond to the percentage of trials on which children advanced the particular answer to the particular problem. For example, an associative strength of 0.02 links the problem 2 – 1 and the answer '0', and an associative strength of 0.87 links 2 – 1 and the answer '1'; these numbers are derived from the children in the overt-strategies-prohibited

Problem	Response 0	1	2	3	4	5	6	7	8	9	10	other
2−1	0.02	0.87	0.05	0.03			0.01	0.01				
3−1	0.01	0.02	0.85	0.01	0.05	0.02		0.02	0.01			
4−1		0.02	0.03	0.86	0.03	0.03		0.01			0.01	
5−1	0.01		0.02	0.04	0.84	0.03	0.03	0.01	0.01			
6−1			0.01		0.05	0.87		0.05	0.01			
3−2	0.03	0.87	0.01	0.01		0.02	0.01	0.03		0.01		
4−2	0.01	0.10	0.71	0.14		0.01		0.03				
5−2	0.01	0.03	0.07	0.74	0.05		0.01	0.03	0.02	0.01	0.01	
6−2	0.01	0.02	0.02	0.11	0.68	0.09	0.02	0.04			0.01	
7−2	0.01	0.06	0.01	0.04	0.05	0.69	0.06	0.01	0.04			0.01
4−3	0.05	0.60	0.20	0.02	0.02	0.05	0.01	0.01	0.02		0.01	
5−3		0.14	0.62	0.06	0.07	0.01	0.02	0.04	0.01	0.02		
6−3	0.01	0.03	0.14	0.55	0.17	0.02		0.03	0.03	0.01		
7−3	0.01	0.01	0.10	0.05	0.55	0.18	0.04		0.03	0.02		
8−3	0.02	0.03	0.03	0.01	0.15	0.53	0.06	0.13		0.02		0.01
5−4	0.09	0.65	0.06	0.03		0.03	0.04	0.03	0.01	0.03	0.02	
6−4	0.02	0.14	0.43	0.18	0.02	0.12	0.02	0.03	0.02	0.01		0.01
7−4	0.01	0.02	0.17	0.43	0.12	0.11	0.04	0.02	0.05	0.01	0.01	0.01
8−4	0.02	0.01	0.09	0.15	0.40	0.12	0.11	0.02	0.04	0.03	0.01	
9−4	0.01		0.01	0.07	0.05	0.54	0.19	0.04	0.06		0.01	
6−5	0.05	0.69	0.06	0.06	0.07		0.02		0.02	0.01		
7−5	0.01	0.10	0.44	0.18	0.13	0.01	0.03		0.07	0.01		0.02
8−5	0.01	0.07	0.07	0.49	0.12	0.05	0.10	0.03	0.01	0.02	0.01	0.01
9−5	0.02	0.03	0.09	0.06	0.53	0.06	0.07	0.05	0.05	0.01	0.01	
10−5	0.01			0.02	0.04	0.73	0.10	0.03	0.01	0.03		0.02

Fig. 5.2. Distribution of associative strengths for kindergarteners' and first-graders' subtraction.

experiment answering 0 on 2 per cent of the trials on this problem and 1 on 87 per cent.

Also as in the Siegler and Shrager (1984) model of addition, the process that operates on this representation can be divided into three phases: retrieval, elaboration of the representation, and counting. As shown in Fig. 5.3, the child (who we here imagine as a girl) first retrieves an answer. If she is sufficiently confident of it, or is unwilling to search further, she states it. Otherwise, she generates a more elaborate representation of the problem, perhaps by putting up and down fingers, and tries again to retrieve an answer. If she is sufficiently confident of the answer, she states it. Otherwise, she counts the objects in the representation and states the last number as the answer.

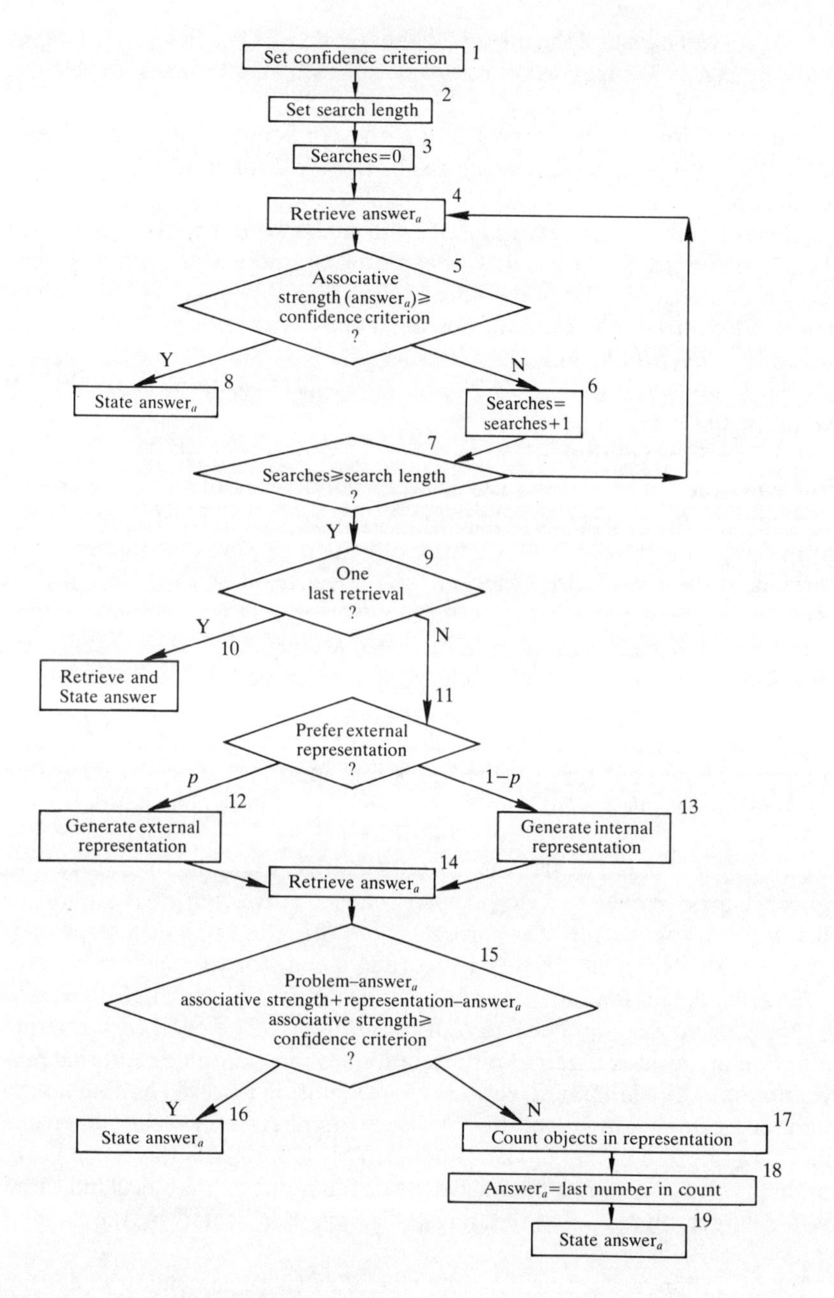

Fig. 5.3. Process for strategy choice in subtraction.

Now we can examine the process in greater detail. The first phase (steps 1 to 10 in Fig. 5.3) involves an effort at retrieval. Each time a problem is presented, the child randomly selects values for two parameters: a confidence criterion and a search length. The confidence criterion defines a value that must be exceeded by the associative strength of a retrieved answer for the child to state that answer. The search length indicates the maximum number of retrieval efforts the child will make before moving on to the second phase of the process. Once these parameters are set, the child retrieves an answer. The probability of any given answer being retrieved on a particular retrieval effort is proportional to the associative strength of that answer for that problem relative to the associative strengths of all answers to that problem. Thus, the probability of retrieving '1' as the answer to '2 – 1' would be 0.87 (Fig. 5.2).

If the associative strength of the retrieved answer exceeds the confidence criterion, the child states that answer. Otherwise, the child examines whether the number of searches that have been conducted is within the permissible search length. If so, the child again retrieves an answer, compares it to the confidence criterion, and advances it as the solution if its associative strength exceeds the criterion. Retrieval efforts continue as long as the associative strength of each retrieved answer is below the confidence criterion and the number of searches does not exceed the search length.

If the point is reached at which the number of searches equals the search length, the child may optionally use an alternate form of retrieval. This form could be labelled the 'sophisticated guessing' approach. It involves lowering the confidence criterion to 0, retrieving one last answer, and stating whichever answer is retrieved. The probability of retrieving any given answer here is again proportional to the answer's associative strength. Intuitively, this process can be likened to trying to spell 'schuss', not having great confidence that any retrieved spelling is correct, and writing out one of the spellings anyway rather than looking up the word in a dictionary or sounding it out.

If the child does not use the sophisticated guessing approach, the next step is to elaborate the representation of the problem. The child can generate either an elaborated external representation or an elaborated internal representation. An elaborated external representation involves putting up the number of fingers indicated by the larger number and putting down the number indicated by the smaller number. An elaborated internal representation involves forming a mental image of the number of objects indicated by the larger number and deleting the number indicated by the smaller number.

Putting up fingers or forming an image adds visual associations between the elaborated representation and various answers to the already existing associations between the problem and various answers. An elaborated external representation involving the child's fingers would add kinesthetic associations as well. I will refer to these visual and kinesthetic associations as 'elaborated representation answer associations' as opposed to the 'problem answer associations' discussed previously. Having formed the elaborated representation, the child again retrieves an answer. If that answer's associative strength exceeds the confidence criterion, the child states it as the answer*. If it does not, the child proceeds to the third phase, an algorithmic process in which he or she counts the remaining objects in the elaborated representation and advances the number assigned to the last object as the difference.

An example of the model's operation

It may be useful to examine how a child using the model would solve a particular problem. Suppose a girl was presented the problem '7 – 4'. Initially, she randomly selects a confidence criterion and a search length. For purpose of illustration, we will assume that she selects the confidence criterion 0.50 and the search length 2. Next, she retrieves an answer. As shown in Fig. 5.2, the probability of retrieving 2 is 0.17, the probability of retrieving 3 is 0.43, the probability of retrieving 4 is 0.12, and so on. Suppose that the child retrieves 4. This answer's associative strength, 0.12, does not exceed the current confidence criterion, 0.50. Therefore, the girl does not state it as the answer. She next checks whether the number of searches has reached the search length. Since it has not, she again retrieves an answer. This time she might retrieve 3. The associative strength of 3, 0.43, does not exceed the confidence criterion, 0.50.

Since the number of searches, 2, has reached the allowed search length, the girl may next use the sophisticated guessing approach in which she would retrieve one final answer from the distribution of associations and state it. If she does not, she next elaborates the problem representation. She does this either by forming a mental image of the initial number of objects and the subsequent decrementing operation or by putting fingers up and down. Suppose that she puts up seven fingers and then puts down four, leaving three fingers up. Next, she again retrieves an answer. Within the parameters of the computer simulation that will be described below, combining the problem–answer and the representation–answer associative strengths increases her probability of retrieving 3 from 0.43 to 0.46. Suppose that she retrieves 3. Its associative strength still does not exceed the 0.50 confidence criterion. Therefore, the girl does not state it. She instead proceeds to the third phase of the process. Here, she counts the fingers that she has up and

* The probability of a given answer being retrieved at this point is determined by adding the problem answer and representation answer associative strengths and dividing by one plus the representation answer associative strength (see Siegler and Shrager (1984, p. 242) for more detail about how this was done).

states the last number as the answer to the problem. If she counts correctly, she will say '3'.

How the model accounts for the data

This model accounts for the strategies that children use, the relative durations of the strategies, and the close relations among the percentage of overt strategy use, the percentage of errors, and the mean solution times on each problem. First consider how it accounts for the existence of the four strategies. The retrieval strategy appears if children retrieve an answer whose problem–answer associative strength exceeds their confidence criterion (steps 1 to 5, sometimes steps 6 and 7, step 8). It also appears when they use the sophisticated guessing approach (steps 1 to 7, 9 to 10). The fingers strategy emerges when children fail to retrieve an answer whose problem–answer associative strength exceeds their confidence criterion, put up their fingers, and then retrieve an answer where the sum of the problem–answer and the elaborated representation–answer associative strengths exceeds their confidence criterion (steps 1 to 7, 9, 11 to 12, 14 to 16). The counting fingers strategy appears if children fail to retrieve an answer whose problem–answer associative strength exceeds their confidence criterion, put up their fingers, fail to retrieve an answer where the sum of the elaborated representation–answer and problem–answer associative strengths exceeds the confidence criterion, and finally count their fingers (steps 1 to 7, 9, 11 to 12, 14 to 15, 17 to 19). The counting strategy is observed if children fail to retrieve an answer whose problem–answer associative strength exceeds their confidence criterion, form an elaborated internal representation, fail to retrieve an answer where the sum of the elaborated representation–answer and problem–answer associative strengths exceeds the confidence criterion, and finally count the objects in the internal representation (steps 1 to 7, 9, 11, 13 to 15, 17 to 19).

The relative solution times of the strategies arise because the faster strategies are component parts of the slower ones. First consider why retrieval should be faster than the fingers strategy. To use the fingers strategy, children must execute all of the steps in the retrieval strategy and five additional ones (four additional ones if the fingers strategy compared to the sophisticated guessing version of retrieval). Next consider why the fingers strategy should be faster than either the counting fingers or the counting strategy. To execute the counting fingers strategy, children must proceed through all of the steps in the fingers strategy and two additional ones. To execute the counting strategy, children must execute all of the steps in the retrieval strategy plus six others. If we can equate the time needed to form elaborated internal and elaborated external representations, children using the counting strategy must execute all of the steps in the fingers strategy plus two others. Thus, the retrieval strategy should be faster than any of the other strategies, the fingers strategy should be faster than the counting fingers strategy, and, if the time needed to form an external representation does not exceed the time needed to form an internal one, the fingers strategy also should be faster than the counting strategy.

Perhaps the most important feature of the model is that it generates close associations among percentage of errors, mean solution time, and percentage of overt strategy use on each problem. The associations arise because all three dependent variables are functions of the same independent variable: the distribution of associations linking problems and answers. The way in which this dependency operates becomes apparent when we compare the outcomes of a peaked distribution of associations, such as that for $4-2$ in Fig. 5.2, with those of a flat distribution, such as that for $7-5$. (A peaked distribution is one in which the largest amount of associative strength is concentrated in a single answer; a flat distribution is one in which several answers have substantial associative strength, though one may have more than the others.)

A lower percentage of use of overt strategies, a lower percentage of errors, and a shorter mean solution time all accompany the peaked distribution. Relative to the flat distribution, the peaked distribution results in (1) less frequent use of overt strategies (because the answer that is retrieved is more likely to have high associative strength, which allows it to exceed more confidence criteria, leading to more use of retrieval and less of overt strategies); (2) fewer errors (because of the higher probability of retrieving and stating the answer that forms the peak of the distribution, which generally will be the correct answer); and (3) shorter solution times (because the probability of retrieving on an early search an answer whose associative strength exceeds any given confidence criterion is greater the more peaked the distribution of associations).

Source of the relations among strategy use, errors, and solution times

The distribution of associations model predicted that the correlations among percentage of overt strategy use on each problem, percentage of errors on that problem, and mean solution time on that problem were due to the patterns of errors and solution times on a subset of trials. Percentage of errors and mean solution times *on retrieval trials*, like percentage of overt strategy use, should depend entirely on the distribution of associations. As described above, on these trials a peaked distribution of associations would lead to fewer errors, shorter solution times, and more use of retrieval rather than an overt strategy. However, errors and solution times on counting fingers and counting trials on a problem should be unaffected by the peakedness of the distribution. Here, children have given up on retrieving a stateable answer from their distribution of associations, so the distribution cannot directly affect relative percentages of errors and solution times on different problems. Instead, the number of counts that children would need to make to execute the strategy on that problem should at least partially determine how long it would take to execute and how many errors it would elicit. If children count by one's, the number of counts on each problem equals twice the size of the larger number. Thus, the model predicts that percentage of overt strategy use should be more highly correlated with errors and solution times on retrieval trials than with errors and solution times on counting fingers and counting trials.

These predictions proved accurate. Percentage of errors on retrieval trials on each problem correlated $r = 0.83$ with percentage of overt strategy use on the problem. In contrast, percentage of errors on counting and counting fingers trials correlated only $r = 0.25$ with percentage of overt strategy use (Fig. 5.4). The difference was highly significant ($t(22) = 4.09$, $p < 0.01$). When the contribution of the size of the larger number was partialed out from both correlations, the difference remained highly significant ($r = 0.72$ vs. $r = 0.00$; $t(22) = 3.51$, $p < 0.01$).

The correlations involving solution times showed a similar pattern, at least when the contribution of the number of counts was partialed out.

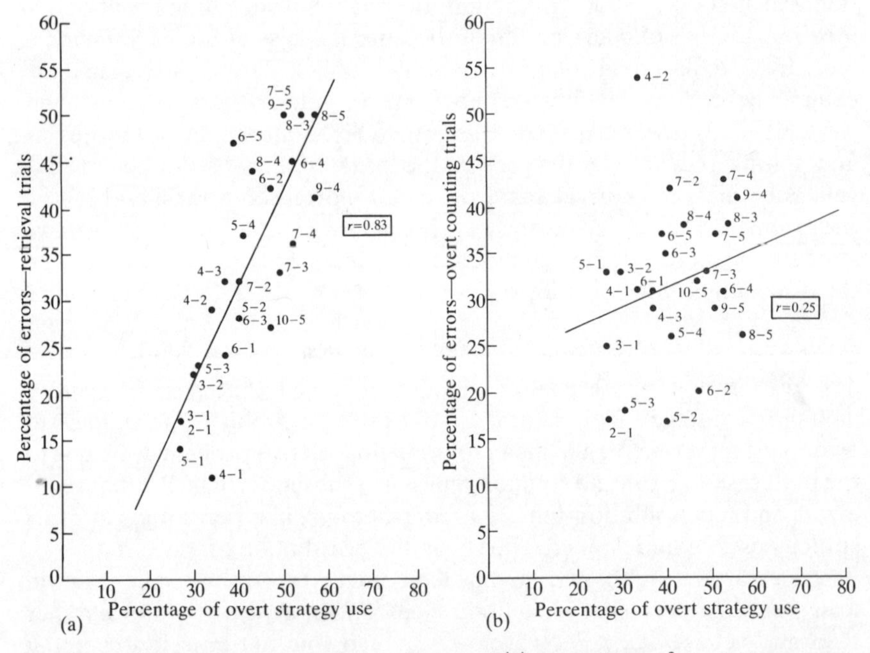

Fig. 5.4. Comparison of correlations between (a) percentage of overt strategy use on each problem and percentage of errors on retrieval trials on that problem and percentage of errors on retrieval trials on that problem and (b) percentage of overt strategy use on each problem and percentage of errors on counting and counting fingers trials on each problem.

Solution times on retrieval trials on each problem correlated $r = 0.83$ with percentage of overt strategy use on the problem; solution times on counting and counting fingers trials correlated with it $r = 0.73$. This difference was in the expected direction, though it was not significant ($t(22) = 1.39, p > 0.05$). The difference between the correlations became significant, however, when the number of counts required on each problem was partialed out ($r = 0.62$ vs. $r = 0.25$; $t(22) = 2.45$, $p < 0.05$).

Thus, the distribution of associations model accurately indicated the type of trial that was responsible for the strong correlations. The source of the correlations of errors and solution times with overt strategy use appeared to be primarily the pattern of performance on retrieval trials which, like the pattern of overt strategy use, reflected the distributions of associations for each problem. This was additional evidence of the usefulness of the model in accounting for young children's subtraction performance. Next, the discussion turns to the way in which the model accounts for development.

How children learn to subtract

The distribution of associations model suggested that learning of subtraction involves both acquisition of problem-solving procedures and acquisition of associations between problems and answers. As discussed by Siegler and Shrager (1984), the problem-solving procedures are probably acquired in a variety of ways. Retrieval seems to be a biologically given capability. The fingers, counting, and counting fingers strategies may be modelled and explained by parents and peers, taught in school, or invented by the children themselves (Groen and Resnick 1977).

The focus of the present discussion will be on the other side of the equation, the development of associations between problems and answers to the problems. Why do some problems come to have peaked distributions and others flat ones? Errors, solution times, and overt strategy use are all attributable to the peakedness of the distribution of associations, but what influences lead to the distributions of associations for different problems assuming the forms that they do?

The basic assumption that the distribution of associations model makes about learning is that children associate whatever answer they state with the problem on which they state it. This assumption reduces the issue of what factors lead children to develop a particular distribution of associations on each problem to the issue of what factors lead children to state particular answers on each problem.

Two classes of factors are assumed to influence which answers children state. One concerns the difficulty of executing the back-up strategies. Factors that increase the difficulty of executing these strategies will lead to a greater number of errors, building associative strength between the problem and incorrect answers and thereby hampering the construction of peaked distributions of associations.

The other class of influences is interference from related numerical operations. The importance of this class of influences is less intuitively obvious, but a number of findings attest to its reality. Both adults and children fail to perfectly separate numerical operations. Winkelman and Schmidt (1974) reported that on verification tasks, adults are slow to reject as false those answers to problems where the answer would be correct for a related numerical operation (e.g. $3 \times 4 = 7$). Miller, Perlmutter, and Keating (1984) reported that for both additional and multiplication, large percentages of adults' errors on a production task resulted from their judging as true those answers that would have been true for another operation. The phenomenon is not limited to interference among different arithmetic operations but rather seems to extend across numerical operations in general. For example, Siegler and Shrager (1984) found that 4- and 5-year-olds' addition performance reflected interference from their knowledge of the counting string. On all problems where the second number was greater than the first, their most frequent error was the number that is one greater than the second number (e.g. $2 + 3 = 4$ and $2 + 4 = 5$)*.

Children's acquisition of peaked or flat distributions of associations on particular subtraction problems would thus be expected to reflect the relative difficulty of executing back-up strategies on them and also the helpful or harmful influence of other numerical operations related to subtraction. Analyses of our best estimate of the 5- and 6-year olds' distributions of associations for subtraction, their answers in the overt-strategies-prohibited experiment (Fig. 5.2), suggested that two factors of the first type and one of the second were particularly influential: the amount of counting down required by the smaller number, whether the problem involved subtracting exactly five objects, and how well learned the corresponding addition problem was.

The amount of counting down

Counting down from the larger number the number of times indicated by the smaller number is among the most widely observed subtraction strategies. It has been observed among children ranging in age from 3 to 13 by Ilg and Ames (1951), Lankford (1974), Starkey and Gelman (1982), Svenson and Hedenborg (1979), and Woods *et al.* (1975), as well as in the present

* Both of these classes of influences, difficulty of executing back-up procedures and interference from other numerical operations, also were present in the Siegler and Shrager (1984) simulation of addition. A third class of influence included there, frequency of presentation of different problems, was not included here, however. The reason was a lack of data concerning presentation rates of these subtraction problems. I plan in the future to examine relative presentation rates of the subtraction problems and to incorporate this variable into the simulation if it proves valuable in accounting for children's subtraction performance.

experiment, Thus, the amount of counting down required by a problem would seem to be a likely contributor to the degree of peakedness of the distribution of associations.

Children sometimes count up to the larger number and/or count up to the difference between the two numbers as well as counting down. Thus, the total number of counts up and down required to solve a problem might be expected to be a better predictor of the peakedness of the distribution of associations for that problem than the smaller number alone. The data indicated that this was not the case, however. The total number of counts that could be made on the problem accounted for 38 per cent of the variance in the associative strengths of the correct answers to the 25 problems. The size of the smaller number accounted for 57 per cent. Further, within the regression analysis described below, the total number of counts required by the problem never added significant independent variance to that which could be accounted for by other factors, among them the size of the smaller number.

The reason that the total number of counts did not influence the development of the distribution of association is probably due to the extremely low error rate using this procedure. More than 80 per cent of the children's errors in counting on the counting fingers trials came when they were counting down rather than up, even though such downward counts were less than 30 per cent of the total counts. This higher error rate seems attributable to both the relative unfamiliarity of counting down and the short-term memory demands imposed by the counting-down process. When counting down, children need to keep track of how many counts down they have made, how many more they need to make, and how much of the larger number is left. When counting up, they only need to keep track of how many counts they have made and whether they have reached the intended stopping point. The amount of counting up required on a problem seems to influence the amount of time needed to execute an overt strategy, but not the number of errors made on overt strategy trials, and therefore not the distribution of associations.

Whether the smaller number equals five

Although counting down the number of times indicated by the smaller number generally became more difficult as the subtrahend increased in size, this was not the case when the subtrahend equalled five. Here, rather than having to decrease the larger number one number at a time, children sometimes put down an entire hand's worth of fingers. The videotapes indicated that the 5- and 6-year-olds used this approach on approximately one-fifth of the trials on which the smaller number was five. They usually answered both quickly and accurately when they did so. Siegler and Robinson (1982) noted a similar phenomenon in 4- and 5-year-olds' addition;

there too, problems where one or both addends were fives were easier than the problem's sum would have suggested.

Influence of corresponding addition problems

Since early in this century, investigators have suspected that addition knowledge might influence the learning of subtraction. Several investigators have found that teaching addition and subtraction problems together aids learning of the subtraction problems (Brownell 1928; Buckingham 1927; Menon 1983). Even when addition and subtraction are not taught together, performance on subtraction problems often correlates highly with performance on inverse addition problems (the difficulty of $5+3=8$ is related to the difficulty of $8-3=5$). Knight and Behrens (1928) reported a correlation of $r=0.64$ between second-graders' knowledge of the 100 inverse addition and subtraction facts were both the first and second addends (the number being subtracted and the difference) ranged from 0 to 9, inclusive. Menon (1983) reported a correlation of $r=0.68$ for first-graders' knowledge of a subset of 64 of these addition and subtraction facts. I used the percentages of correct answers that Starkey and Gelman (1982, p. 103) reported for a set of 15 corresponding addition and subtraction problems where the sum in the addition problems and the larger number in the subtraction problems did not exceed 6. The correlation between the 4-year-olds' percentage correct on the corresponding addition and subtraction problems was $r=0.83$; the correlation for the 5-year-olds was $r=0.76$.

How might knowledge of addition influence subtraction performance? The following path seemed to represent one plausible way. Suppose that children were asked 'Eight minus five equals what?' They might translate the problem to the form 'What plus five equals eight?' Children who had a very peaked distribution for $3+5=8$ would presumably be more likely to retrieve the answer 3. In general, the more peaked the distribution of associations for the inverse addition problem (for $A-B=C$, the inverse addition problem would be $C+B=A$), the easier the problem should be.

A test of the model of development of the distribution of associations

To test the adequacy of this account for development, I qualified each of the three hypothesized predictor variables and examined how well they together accounted for the associative strengths of the correct answers in the Fig. 5.2 distribution of associations. One predictor variable was the size of the smaller number. A second predictor was whether the smaller number equalled five: this variable was quantified by coding it as 1 if it equalled five and as 0 if it did not. A third predictor was the associative strength of the correct answer to the exact inverse addition problem. Values for this variable were the associative strengths of the correct answers to the 25 addition problems reported by Siegler and Shrager (1984, p. 240). Illustratively, for the problem $5 - 4$, the value of this predictor was 0.61, because this was the associative strength linking the problem $1 + 4$ and the answer 5.

The regression analysis supported the above analysis of development in several ways. The three predictor variables accounted for 84 per cent of the variance in the associative strengths of the correct answers on the 25 problems in Fig. 5.2. Each of the predictors added significant amounts of variance to that which could be explained by the other two. The associative strength of the corresponding addition problem was the first to enter the equation, accounting for 64 per cent of the variance; the size of the smaller number was the next to enter, adding 7 per cent; and whether the smaller number was five added an additional 13 per cent to the variance that could be accounted for. Removing any one of the variables from the equation would have significantly lowered the percentage of variance that could be explained. Several other variables did not enter the equation at any point and could not account for additional significant variance: the size of the larger number, the size of the sum of the larger and smaller numbers, and the difference between the larger and smaller numbers. Neither the Woods *et al.* (1975) nor the Svenson and Hedenborg (1979) models accounted for nearly as much variance alone, nor could they enter into the equation at any point. The Woods *et al.* model accounted for 34 per cent of the variance and the Svenson and Hedenborg model for 22 per cent*.

These results were consistent with the view that the size of the smaller number, whether five was the number being subtracted, and the associative strength of the corresponding addition problem contributed to the development of associations between subtraction problems and their answers. More generally, they suggested that, in subtraction as in addition, difficulty of executing back-up strategies and the influence of related operations are two of the factors influencing the relative difficulty of problems.

A computer simulation of performance and learning of subtraction

The above analysis allowed formulation of a computer simulation of subtraction that both performs and learns. The model progresses from producing relatively poor performance at the outset to eventually producing the quite sophisticated performance typical of 5- and 6-year-olds. (The simula-

* The reason for the Svenson and Hedenborg model accounting for less variance than the Woods *et al.* model was that, for these kindergarteners and first-graders, problems where the larger number was exactly twice as great as the smaller were at least as difficult as their smaller numbers would indicate. These problems apparently become much easier at somewhat older ages than the ones examined here, since Woods *et al.* and Svenson and Hedenborg reported that second- to fifth-graders solve them much more quickly than their numerical sizes alone would suggest.

tion is written in Lisp, a common simulation language. The code was written by Chris Shipley, and is available upon request.)

An outline of the simulation

The simulation can be described in terms of its representation and process at the outset and in terms of nine features of its operation. At the outset, the representation includes only two types of knowledge. One is the understanding that numbers as a class are appropriate answers to subtraction problems. This information is represented as a set of minimal associations (associative strength = 0.01) between each problem under consideration and each possible answer (each whole number between 1 and 11). Second, when a problem is presented, the association between that problem and its correct answer is strengthened for the duration of the trial by an amount proportional to the association between the exact inverse addition problem and its answer (when $4 - 1$ is presented, the answer 3 is temporarily strengthened by an amount proportional to the associative strength linking $3 + 1$ and 4).

Now consider the initial state of the process. Biology has given us the ability to retrieve information from memory. Direct instruction and modelling have taught us to represent and count the numbers in the problem. Thus, the process at the beginning of the simulation resembles that shown in Fig. 5.3.

This initial representation and process are insufficient to produce performance as advanced as that of 5- and 6-year-olds. For example, when we presented to the simulation in its initial state a set of 10 000 problems, 400 each of the 25 problems in Fig. 5.2, it answered correctly on only 53 per cent of trials and stated a retrieved answer on only 5 per cent of trials. In contrast, the children answered 73 per cent of problems correctly and stated retrieved answers on 58 per cent of trials.

What the initial representation and process do provide is a base from which learning can occur. The acquisition process can be described in terms of nine features:

(1) The simulation is presented each of the 25 problems 80 times.

(2) Before each trial, the simulation generates a confidence criterion and a search length. To obtain the most representative run possible, each of the confidence criteria between 0.05 and 0.95 is selected equally often and each search length is selected equally often (subject to rounding).

(3) The probability of retrieving an answer is proportional to its associative strength compared to the associative strengths of all answers to the problem. A retrieved answer is stated if its associative strength exceeds the current confidence criterion. Retrieval attempts continue until either the associative strength of a retrieved answer exceeds the confidence criterion or the number of searches matches the allowed search length.

(4) On 20 per cent of trials where the number of searches has reached the allowed search length and no answer has been stated, the confidence criterion is reset to 0 and one last answer is retrieved and stated. The probability that a given answer will be stated is, as above, proportional to its associative strength relative to all answers' associative strengths.

(5) If no answer has been stated, the program generates an elaborated representation of the number of objects in the larger number (the minuend). On 75 per cent of trials, the elaboration involves putting up fingers. On 25 per cent of trials, the elaboration involves forming a mental image of the specified number of objects.

(6) The program next decrements the larger number by the number of objects indicated in the smaller number (the subtrahend). On 20 per cent of the trials where the smaller number equals five and where there is an external representation, the simulation recognizes the possibility of putting down one full hand's worth of fingers and correctly does so. On all other trials, the model deletes the number of objects indicated in the subtrahend, counting them one by one. On each downward count, there is a fixed probability of skipping over the object being counted and a fixed probability of counting it twice.

(7) Once the objects in the smaller number have been deleted, the model temporarily (for the duration of the trial) adds a constant to the associative strength of the answer corresponding to the number of objects that are left. In accord with the convention established by Siegler and Shrager (1984), the constant value that is added is 0.05. The program then retrieves an answer, and states it if the answer's associative strength exceeds the confidence criterion.

(8) If the associative strength does not exceed the confidence criterion, the program counts the remaining objects. The last number counted is stated as the answer.

(9) Every time the system advances an answer, the association between that answer and the problem increases. The increment is higher for correct answers, that presumably are reinforced, than for incorrect answers, that presumably are not*.

The simulation's behaviour

The simulation runs in two phases: a learning phase and a test phase. The only difference between the two phases involves the last point listed above.

*This description differs from the Fig. 5.3 flow diagram both in including numerical values for several probabilities and in specifying learning features local to subtraction, such as when subtracting five is performed through closing a full hand's worth of fingers. The difference can be viewed as in Fig. 5.3, which indicates the framework of the model that is provided by the general distributions of associations approach, with the computer simulation adding to that framework particular parameter values and particular customizations local to the case of subtraction.

During the learning phase, each answer adds strength to the association between that problem and that answer. During the test phase, no strength is added. The reason is that the test phase is intended to model the experimental situation, in which many children are presented each problem one or two times, rather than a single child performing each problem many times.

The simulation's performance during the test phase proved to closely resemble that of the 5- and 6-year-olds who were observed. It generated all four strategies that the children used and no others. The relative solution times of the simulation's strategies also were identical to those produced by the children. The absolute percentage of correct answers produced by the simulation, 80 per cent, was similar to that of the children, 73 per cent, as was its percentage of overt strategy use (34 per cent vs. the children's 42 per cent).

Most important, the simulation's performance resembled that of the children in which problems elicited the greatest percentage of errors, which took the longest to answer, and which elicited the highest percentage of overt strategies, as well as in producing high correlations among the three variables. As shown in Table 5.2, all of the correlations of greatest interest between the simulation's behaviour and that of the children equalled or exceeded $r = 0.80$. Moreover, the intrasimulation correlations among percentage of errors on retrieval trials, percentage of overt strategy use, and mean solution times on retrieval trials on each problem all equalled or exceeded $r = 0.95$.

Table 5.2. *Computer simulation's performance in test phase.*

Intramodel correlations	Correlations between children's and model's behaviour
$r_{\%}$ errors and % overt strategy use = 0.96	$r_{\%}$ errors by model and children = 0.80
$r_{\%}$ errors and $\bar{x}$ solution times = 0.95	$r_{\%}$ overt strategy use produced by model and children = 0.81
$r_{\%}$ overt strategy use and $\bar{x}$ solution times = 0.96	$r_{\bar{x}}$ solution times produced by model and children = 0.84

In sum, a computer simulation that takes into account the size of the smaller number, whether that smaller number equals five, and the strength of the corresponding addition problem can produce subtraction performance much like that of 5- and 6-year-olds. The simulation both performs and learns. At the outset, its perfomance is not very accurate, and is unlike the reasonably proficient performance of 4- and 5-year-olds in many ways. After having an opportunity to learn, its performance is much more childlike. The simulation demonstrates that children could acquire their distributions of associations through the three hypothesized mechanisms and that, if they did, their performance would be much like that which was observed.

Conclusions

In closing, I would like to note three general advantages of depicting within a single model the diverse problem-solving strategies that children use to solve particular problems. First, only by depicting within single models the multiple strategies that children use in many domains can the models accurately describe what children do to solve such problems. Second, models that account for the diversity of children's strategies may also prove capable of accounting for many other aspects of their performance. For example, the distribution of associations model accounted not only for the several strategies children used to solve subtraction problems, but also for the relative solution times of the strategies, the relative error rates on different problems, the relative solution times on different problems, the relative frequencies of use of overt strategies on different problems, and the high correlations among errors, solution times, and overt strategy use. Not considering the variety of strategies children use forces averaging over different approaches, which inherently limits the range of data that can be explained and the precision with which it can be explained. Third, models that account for the diversity of children's strategies may also help to integrate learning with performance. Within the present model of subtraction, as well as in the Siegler and Shrager (1984) model of addition, learning is largely a function of the factors that make performance via different strategies easy or difficult. By considering what strategies children use to solve particular problems and what makes performance easy or difficult when they use each strategy, we may gain a better understanding of why learning takes the form it does.

References

Brownell, W. A. (1928). *The development of children's number ideas in the primary grades*. University of Chicago Press, Chicago, IL.

Buckingham, B. R. (1927). Teaching addition and subtraction facts together or separately. *Educational Research Bulletin* **6,** 228–9, 240–2.

Groen, G. J. and Resnick, L. B. (1977). Can preschool children invent addition algorithms? *Journal of Educational Psychology* **69,** 645–52.

Ilg, F. and Ames, L. B. (1951). Developmental trends in arithmetic. *The Journal of Genetic Psychology* **79,** 3–28.

Knight, F. B. and Behrens, M. S. (1928). *The learning of the 100 addition combinations and the 100 subtraction combinations*. Longmans, Green and Co., New York.

Lankford, F. G. (1974). What can a teacher learn about a pupil's thinking through oral interviews? *The Arithmetic Teacher* **21,** 26–32.

Menon, R. (1983). *Teaching basic subtraction facts to ten using an additive strategy*. Masters' Thesis, University of Northern Iowa, Ceder Falls, IA.

Miller, K., Perlmutter, M. and Keating, D. (1984). Cognitive arithmetic: comparison of operations. *Journal of Experimental Psychology: Learning, Memory, and Cognition* **10,** 46–60.

Siegler, R. S. (in press). Unities in strategy choices across domains. In *Minnesota symposium on child development Vol. 21*. (ed. M. Perlmutter). University of Minnesota Press, Minneapolis.

Siegler, R. S. and Robinson, M. (1982). The development of numerical understandings. In *Advances in child development and behaviour Vol. 16*. (eds. H. Reese and L. P. Lipsitt), Academic Press, New York.

Siegler, R. S. and Shrager, J. (1984). Strategy choices in addition and subtraction: how do children know what to do? In *The origins of cognitive skills* (ed. C. Sophian), Erlbaum, Hillsdale, NJ.

Siegler, R. S. and Taraban, R. (1986). Conditions of applicability of a strategy choice model. *Cognitive Development* **1,** 31–51

Starkey, P. and Gelman, R. (1982). The development of addition and subtraction abilities prior to formal schooling in arithmetic. In *Addition and subtraction: a cognitive perspective*. (eds. T. P. Carpenter, J. M. Moser, and T. A. Romberg, Erlbaum, Hillsdale, NJ.

Svenson, O. and Hedenborg, M. (1979). Strategies used by children when solving simple subtractions. *Acta Physchologica* **43,** 477–89.

Winkelman, J. and Schmidt, J. (1974). Associative confusions in mental arithmetic. *Journal of Experimental Psychology* **102,** 734–6.

Woods, S. S:, Resnick, L. B. and Groen, G. J. (1975). An experimental test of five process models for subtraction. *Journal of Educational Psychology* **67,** 17–21.

6

The role of associative interference in learning and retrieving arithmetic facts

JAMIE I. D. CAMPBELL

Most adults take for granted their knowledge of simple number facts like $5+8$ or 6×9. It is easy to forget the five or six years of classroom drill that was required to establish this fundamental skill. One may be surprised in retrospect, however, by the difficulty of learning the simple combinations. Consider the multiplication combinations ranging from 0×0 to 9×9. All of the zero-times and one-times problems can be solved by recourse to two simple rules (i.e. $N \times 0 = 0$ and $N \times 1 = N$). Among the remaining problems from 2×2 to 9×9 there are only 36 problems when operand order is ignored (i.e. when it is realized that 3×4 and 4×3 can be treated as one problem). Campbell and Graham (1985; see also Graham, this volume) found, however, that even after four years of drill in public school, children made errors on 17 per cent of all trials involving these problems. Under only moderate speed pressure a substantial subset of problems produced error rates in excess of 30 per cent. During this same four-year period these children undoubtedly acquired thousands of pieces of new information (e.g. words, facts, concepts, rules). Despite intensive training, however, mastery of the basic multiplication facts remained elusive.

Indeed, it appears that complete mastery may be a rare occurrence. Campbell and Graham (1985) also found that even among university students, error rates on specific combinations (e.g. 4×8, 6×9) approached 30 per cent. These high error rates were observed under a level of speed pressure where a subset of the problems yielded almost no errors at all. Why is memorization of arithmetic combinations so difficult, and why are certain combinations so much more difficult than others? This paper claims that the available evidence supports an associative-network theory of arithmetic fact skill. It is argued that the protracted course of learning, and differential problem difficulty, are due substantially to associative interference. Interference in mental computation provides a powerful experimental tool for exploring semantic memory, and the interference theory has direct implications for the pedagogy of arithmetic.

The problem-size effect

One of the basic features of arithmetic memory is the *problem-size* effect. This refers to the general trend for response times (RTs) to increase as the numerical magnitude of the problem increases (see Campbell and Graham 1985; Miller, Perlmutter and Keating 1984; Norem 1928; Parkman 1972; Stazyk, Ashcraft and Hamann 1982). Figure 6.1 shows RT in milliseconds (msec) for the two-times through nine-times multiplication tables. These data come from 16 adult subjects who generated answers to each problem 12 times. The mean correct RT for problems that have 2 as an operand was about 730 msec whereas problems involving 9 as a multiplier required 975 msec on average. The general problem-size effect also holds for adult's addition (Ashcraft and Battaglia 1978; Ashcraft and Stazyk 1981; Groen and Parkman 1972; Matthews 1985) and simple division (Campbell 1985).

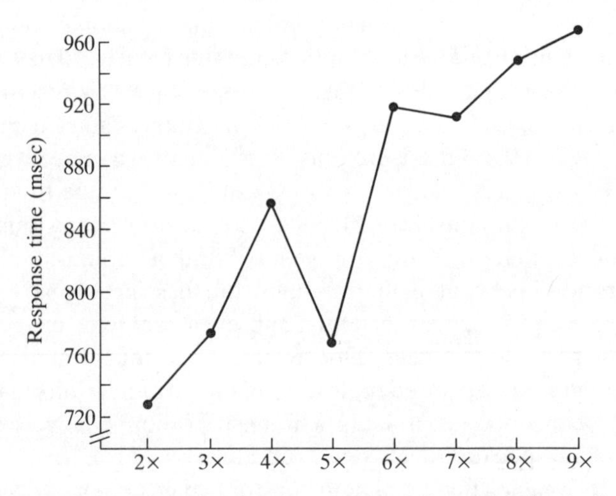

Fig. 6.1. Mean RT in msec for the times-tables 2 through 9 (excluding problems involving 0 or 1).

The problem-size effect in adults has given rise to several theoretical models of arithmetic fact skill. One view maintains that the effect reflects the execution of rule-based procedures. Large-number problems may take longer because they require more computational operations. There is good evidence that children initially use rule-based strategies to perform simple arithmetic (e.g. Ashcraft and Fierman 1982; Siegler and Shrager 1984; Suppes and Groen 1967), and a number of theorists (e.g. Baroody 1983; Parkman and Groen 1971) have suggested that procedural methods may be preserved in adult performance. Advanced skill with the arithmetic facts may involve automatic computational procedures controlled by the numerical values of operands in each problem (e.g. Findlay and Roberts 1985).

Other researchers have argued for network-distance models of arithmetic memory (e.g. Ashcraft 1982, 1983; Ashcraft and Battaglia 1978; Ashcraft and Stazyk 1981; Miller *et al.* 1984; Parkman 1972; Stazyk *et al.* 1982). Encoding a problem causes nodes in the network (i.e. answers to problems) to be available if they are activated sufficiently. The problem-size effect in adults is supposed to derive largely from the search 'distance' in the network. Distance in this context is analogous to the concept of 'trace strength' (Anderson 1983; Ashcraft, Fierman and Bartolotta 1984; Wickelgren 1976), where strength controls the speed and probability of retrieval. Answers to small-number problems are retrieved quickly because their operands are closely associated with the correct answer node in the memory structure. Larger problems take longer because the correct problem-response traces are relatively weak.

Both automatic-computation and network-distance retrieval models are supported by correlations between RT and numerical quantities in each problem, so-called *structural* variables. For example, in multiplication, correlations about 0.6 to 0.8 are found between a problem's RT and the magnitude of the correct product (Campbell and Graham 1985; Miller *et al.* 1984; Stazyk *et al.* 1982). Reliable correlations are also found between RT and the minimum or maximum operand, or the sum or squared sum of the operands (e.g. Stazyk *et al.* 1982). In procedural models the values of structural variables estimate the number of operations or steps executed (e.g. the number of successive increments in a counting procedure). In network-distance models the structural predictors index the organization of strength relationships in the network, an organization that preserves the numerical magnitude of problems.

There are reasons to question the assumption that correlations with structural variables directly reflect the memory process involved. Tie problems (e.g. $2+2$, 6×6, etc.) do not obey a general problem-size rule. Ties are often analysed separately because the slopes of their problem-size functions are lower than for the non-tie problems. Similarly, as Fig. 6.1 shows, problems involving 5 are faster than their numerical magnitude would predict. There are other notable exceptions. For example, 8×9 (or 9×8) should be one of the slowest problems, but Campbell and Graham (1985) found that it failed to rank even in the slowest 25 per cent of problems.

To accommodate exceptions to the simple problem-size rule, an accessibility factor has been proposed that can vary independently of problem-size. Miller *et al.* (1984) distinguished between the location of information in a representation (i.e. network distance which is indexed directly by structural variables) and accessibility of the information. The latter is thought to reflect variations in practice or familiarity (Miller *et al.* 1984; Stazyk *et al.* 1982), but no explicit model of accessibility has been offered in the context of adult arithmetic research. Given, however, that a substantial percentage of problems deviate from problem-size predictions, a specific model of accessibility is clearly in order. Indeed, it will be argued that variations in accessibility are largely responsible for the problem-size effect.

The relationship between errors and retrieval time

How might we begin to formalize the role of accessibility in arithmetic memory? Some of the early research on the development of children's arithmetic skill provides some interesting clues. In a 1928 master's thesis, Grant Norem (see also Norem and Knight 1930) traced learning of the simple multiplication facts for 25 third-grade students. He found highly systematic patterns in the errors made by these children. 91 per cent of the 5400 errors recorded were correct responses to other multiplication combinations. Of these, about 70 per cent were correct answers to 'closely allied combinations', usually ones with a multiplier in common (i.e. from within the same times-table). This is not an isolated finding. Campbell and Graham (1985) observed the development of this pattern in the multiplication performance of children in grades 2 through 5 (see also Graham, this volume). The proportion of errors that were table-related (e.g. $4 \times 6 = 28$) exceeded chance for all groups and increased steadily across grades.

Adults' errors on simple multiplication are also systematic. Campbell and Graham (1985) analysed the errors made by 60 adult subjects. 93 per cent of the 660 errors observed were answers to other problems in the times-tables and, of these, 85 per cent were table-related. The high percentage of table-related errors suggests that most errors are due to associative intrusions. Strong (e.g. correct) associations established in the context of one problem (e.g. $4 \times 6 = 24$ or $3 \times 8 = 24$) promote errors in the context of other problems ($4 \times 8 = 24$). This implies that differential problem difficulty can be described in terms of differential susceptibility to associative intrusions.

The relevance of errors to an understanding of the problem-size effect is quite clear when the relationship between problem-error rate and correct retrieval time is examined. Figure 6.2 presents data for a group of 20 adult subjects that was tested 10 times on each of the simple multiplication problems from 2×2 to 9×9. The correlation between percent errors and correct RT across problems was 0.93. Under instructions for speed, time for a correct retrieval is accurately predicted by the probability of generating an error. The relationship between errors and correct RT across problems is not an artifact of averaging over subjects. With a sufficient number of trials for each problem, a reliable correlation across problems between RT and errors is observed within individuals (Campbell 1985; Matthews 1985).

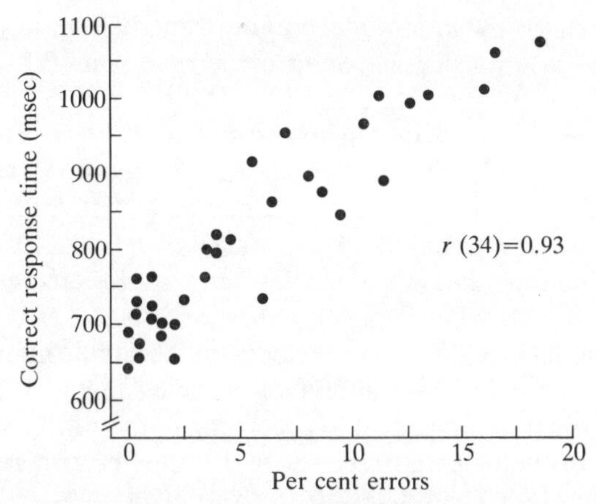

Fig. 6.2. Mean correct RT plotted against per cent errors for the 36 multiplication problems ranging from 2 × 2 to 9 × 9.

The relationship between problem-error rate and retrieval time emerges early in the learning process. Campbell and Graham (1985) found for each of the four school grade-levels tested, and for adult subjects, that both problem-error rate and product-error rate (how often a problem's correct product appeared as an incorrect response to other problems) yielded highly significant correlations with correct RT for problems. Both error measures always contributed independently to predicting RT in multiple regression analyses, yielding an average multiple r of 0.83. The error variables always yielded higher correlations than did any of the standard structural variables. Surely, the systematic nature of errors, and the strong correlations between problem-errors, product-errors, and correct RT reveals some basic structural aspect of memory for multiplication facts.

Promiscuous connections and network interference

Norem (1928) made several other interesting observations about children's multiplication skill. Although he found that the magnitude of a problem's correct product was a good predictor of difficulty, he also suggested that a problem was difficult to learn if its correct product was susceptible to what he called 'promiscuous connection forming' (Norem 1928, p. 29). More specifically, he found that the responses most often occurring as errors were frequently correct products to problems requiring the most learning trials. Norem claimed that the relationship between learning difficulty and product-error rates indicated that children established false associations during learning that weaken correct associations or interfere with the formation of correct associations.

Casting Norem's view in a modern light, Campbell and Graham (1985) argued that the relationship between error rates and RT, and between product-error rate and learning difficulty, reflects a process of interference among candidate answers stored in an associative network. They coined the term *arithmecon* to refer to the hypothetical network. According to the network-interference theory, the frequency of a particular error can be taken to indicate the strength of an associative trace (or promiscuous connection if you like) linking a problem and a false answer. Siegler and Shrager (1984) similarly have conceptualized errors to define the 'distribution of associations' in children's knowledge of the simple number combinations. Each problem can be thought of as being linked to a number of candidate answers that are activated according to their trace strengths. The strong correlation between correct RT and problem-error rate occurs because, even when false candidates are not retrieved, their activation nevertheless slows correct retrieval time. Thus, according to the interference account, differential accessibility of correct facts largely reflects differing amounts of interference caused by the activation of multiple candidates that compete for retrieval (cf. Anderson 1981, 1983; Campbell and Graham 1985; Siegler and Shrager 1984). The fact that product-error rates contribute independently of problem-error rates in multiple regressions predicting correct RT suggests that false associations with a problem's correct answer are an additional source of interference.

Before considering how associative interference could give rise to the problem-size effect (i.e. why interference should be more pronounced for large-number problems), additional evidence for interference and an associative candidate structure should be considered. So far, the argument for interference has rested solely on the systematic nature of errors and on the correlation between error rates and correct RT across problems.

Manipulating interference in mental multiplication

To investigate associative interference in simple multiplication, Campbell (1985) took advantage of two other phenomena. First, RTs become faster with repeated retrievals (e.g. Stazyk *et al.* 1982). It is commonly held in network theories that speed-up with repetition reflects strengthening of the link or reducing the length of the pathway to a retrieval node (Anderson 1981, 1983; Stazyk *et al.* 1982). According to the network-interference model (Campbell 1985; Campbell and Graham 1985), however, the speed of correct retrieval also depends upon the concurrent activation level of other associated candidate products. Specifically, for a particular problem, when the activation levels of false candidates are high, a correct RT will be slower than when false candidates are relatively quiescent. If this is correct, the course of speed-up with practice is also a function of the amount of interference imposed by false associations at the moment of each practical trial; for example, the speeding effect of repetition could be amplified if competitive interference in the memory structure was reduced.

A second effect found in simple multiplication experiments suggests a method for manipulating interference at the level of individual problems. Campbell (1984, 1985) called the effect *error priming*. This refers to the fact that retrieval of a product increases the probability that this answer will occur as an error response on a subsequent trial. When an error occurs, the probability that the same answer was given (usually as a correct product) on a recent prior trial is approximately 10 to 20 per cent higher than chance would predict.

Campbell (1984, 1985) argued that error priming implies a direct relationship between correct and erroneous retrievals of the same product. Retrieval of an answer via one problem temporarily produces a high level of activation at that answer node in the arithmecon. If another problem is subsequently encountered that also accesses that candidate, the activation resulting from the current trial combines with the residual activation from the prior encoding episode, thereby increasing the likelihood of retrieving that answer.

If the relationship between errors and RT in simple arithmetic reflects an interference process that controls both, then it follows from this explanation of error priming that prior retrieval of a problem's false-candidate answers should impede a subsequent correct retrieval, because the residual activation of false candidates competes with the activation at the correct answer. Conversely, eliminating retrieval of false candidates should facilitate RT for associated problems, relative to trials where residual activation has temporarily amplified interference. Thus, it should be possible to affect the course of speed-up with repeated retrievals by manipulating the particular problems a subject encounters.

Campbell (1985) tested these predictions by giving adult subjects initial practice on all 36 problems from 2×2 to 9×9. Then, a specific subset of 18 problems, the interference set, was withdrawn while the subjects continued to practice the remaining combinations. Recall that about 90 per cent of multiplication errors involve answers to other multiplication problems. Error patterns observed in earlier experiments were used to assign problems to the practice and interference sets so that the majority of typical errors on the practice problems involved answers to problems in the interference set. Such errors are referred to as *between-set* errors. The assumption was that the frequency of specific errors can be used to identify the false candidate answers most strongly associated with each problem (cf. Siegler and Shrager 1984). The purpose was to have the strongest false candidates for the practice set as correct products to interference problems.

There were 13 subjects who were tested for 10 sessions, five sessions on each of two consecutive days. In the first session the subjects generated answers to all 36 problems, once in each of four randomized trial blocks. However, during sessions two through nine, only the 18 problems in the practice set were tested, again, four times in each session. In the tenth session the interference set was reintroduced and tested in random order among the practice-set problems.

The network-interference theory makes the following predictions. In Session 1, when the interference set is tested among the practice-set problems, retrieval of interference-set products should cause high levels of activation at these response nodes in the network. Because the interference-set products are, on average, strong false associates of practice-set problems, this activation should disrupt practice-set performance. The interference should be expressed as a high proportion of between-set errors on the practice-set problems.

When the interference problems are withdrawn in Session 2, activation of their product nodes should begin to decay, and the probability of these answers occurring as errors on practice problems will decrease, that is, error-priming of these answers should be eliminated, resulting in a decline in the rate of between-set errors. If the activation levels of interference-set products affect the speed of correct retrieval for practice-set problems, then the shape of the RT function across practice sessions will follow the decay function of interference-set product activation. Thus, across sessions there should be a high correlation between practice-set RT and the frequency of between-set errors. Finally, reintroducing the interference problems in Session 10 should reactivate their product nodes, increase the probability of their erroneous retrieval, and slow RT for the practice set.

Figure 6.3 shows mean RT for the 18 practice-set problems plotted over the ten sessions, as well as total errors made on the practice problems across sessions. The overall error rate for practice-set problems was 4 per cent. About 65 per cent of the practice-set errors in Session 1 involved correct products to interference-set problems; that is, between-set errors. The hatched area in the bars in Fig. 6.3 represents the number of between-set errors that occurred on practice problems in each session.

It is clear that there was a systematic decline in between-set errors when the interference problems were withdrawn (65 per cent of errors in Session 1 to 10 per cent of errors in Session 9), an effect that reflects the elimination of priming of interference-set products.

Over the first nine sessions the correlation between practice-set RT and the number of between-set errors was 0.95. The same correlation for total errors was 0.75. Clearly, speed-up over sessions was closely tied to the probability of between-set errors. In Session 10, when the interference problems were reintroduced, between-set errors on practice problems reappeared, accounting for 59 per cent of all errors. As predicted, there

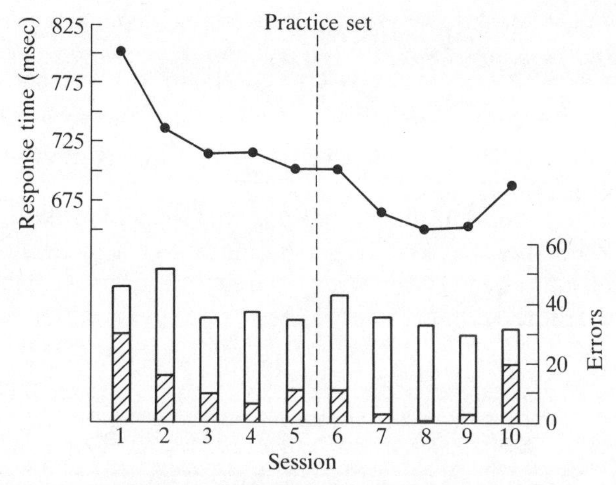

Fig. 6.3. Mean correct RT for the 18 problems in the practice set plotted over sessions. Each bar indicates the total number of errors made on practice problems in the corresponding session. The hatched area represents the proportion of between-set errors.

was a significant 29 msec increase in practice-set RT from Session 9 to Session 10 that corresponded to the reappearance of between-set errors.

To confirm problem-specific interference, Campbell (1985) repeated the experiment with the practice-set problems selected to define a control practice set and an experimental practice set. The two sets involved 12 problems each and the sets were distinguished by their relationship to the remaining 12 problems which constituted the interference set. The control practice set was selected so that only a small proportion of the common errors for these problems (about 14 per cent) were between-set errors; that is, errors involving answers to the interference set. However, between-set errors accounted for about 65 per cent of the errors typically observed for the problems assigned to the experimental practice set. The two practice sets were also balanced for mean RT and overall error rates.

In this experiment, 14 subjects were tested for five mini-sessions on a single day. In the first session all 36 problems were tested once in each of four trial blocks. Problem order was random, with the exception of the interference set which always occupied the first 12 trials in a block. In each of the middle three sessions, only the 24 problems in the two practice sets were tested. Each problem was tested four times in each session. In Session 5 the interference set was reintroduced and tested with the practice problems as in Session 1.

The predictions with respect to the two practice sets were as follows. The experimental set should speed-up more than the control set when the interference problems are withdrawn. This follows from the network-interference account because, as indicated by error patterns, products from the interference set are more strongly associated with the experimental problems than with the control problems. Consequently, the decay of activation of interference-set answers, following withdrawal of the interference problems, should primarily benefit the experimental practice set by reducing associative competition in the structure that is specific to those problems. Similarly, when the interference set is reintroduced in Session 5, there should be substantial interference with the experimental set but little or no interference with the control set.

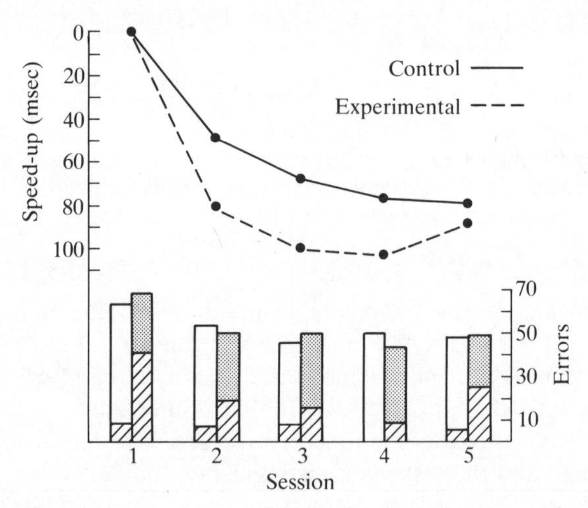

Fig. 6.4. The change in mean correct RT for the control and experimental practice sets over sessions. The unshaded bars represent total errors on the control set and the shaded bars represent experimental set errors. The hatched area in each bar represents the proportion of between-set errors.

Figure 6.4 shows mean RT for the control and experimental practice problems plotted over the five sessions, as well as errors made on the practrice problems across sessions. The a priori assignment of problems to the control and experimental practice sets was verified. Thus, in Session 1, when the interference problems were present, between-set errors constituted 58 per cent of errors on the experimental set but only 13 per cent of control set errors. There was a 7.7 per cent error rate for practice problems across all sessions. Between-set errors are indicated by hatched areas in each bar.

As in the last experiment, the frequency of between-set errors declined when the interference set was withdrawn. By Session 4, between-set errors accounted for only 18 per cent of the experimental set errors and there were no between-set errors on control practice problems. In Session 5, when the interference problems were reintroduced, between-set errors returned, constituting 62 per cent of the errors on the experimental problems, indicating that error-priming of interference-set products was reestablished.

The pattern of RTs supports a theory of problem-specific interference. Over the first two sessions there was a reliable Session × Practice-set interaction such that the experimental set sped-up 28 msec more than the control set when the interference problems were withdrawn. There was no hint of this interaction when the same analysis was performed across half-sessions within Session 1. When the interference problems were present, the rate of speed-up was identical for the control-practice and experimental-practice problem sets. Furthermore, when the interference set was reinserted in Session 5, there was a significant 19 msec increase in experimental set RTs, but there was no effect on RT for the control problems.

These results indicate that simple multiplication problems access a common network structure of candidate responses. Encoding a problem activates multiple response candidates in the network, and different problems can activate intersecting network substructures. Temporally-close activation episodes are not independent. When intersection occurs, activation of a network response node by retrieval creates interference that disrupts performance on other problems that have associative links to that response.

Network interference and the problem-size effect

To summarize, Campbell's (1985) experiments showed that individual multiplication problems activate a network substructure of candidate responses. The activation levels of false candidates affect both the probability of an error and the speed of a correct response. Priming a response by a prior retrieval promotes errors and slows correct RT only for problems that have a relatively high probability of generating that product as an error. This indicates a direct relationship between speed of correct retrieval and the strength of false associations as measured by the frequency of particular errors. The results of Campbell's (1985) practice/interference experiments support the theory (Campbell and Graham 1985; Graham, this volume) that the strong correlation between error rates and correct RT across problems reflects an associative interference process that controls both. Since the error variables are better predictors of correct RT than structural variables (Campbell and Graham 1985; Graham, this volume), it follows that the problem-size effect is also due substantially to a process of associative interference.

If the effect is due mainly to interference, however, why should interference have a greater impact on large-number problems than on small-number problems? One probable factor is that small-number combinations are practiced more frequently than large-number combinations (Campbell and Graham 1985; Stazyk *et al.* 1982). It has long been held that frequency of occurrence is supposed to determine the strength of associations (e.g. Anderson 1983; Thorndike 1922). A strong correct association would be less susceptible to interference than a weak correct association. For the four basic arithmetic operations, Clapp (1924) found respectable correlations (averaging about -0.4) between a problem's difficulty ranking based on children's error rates and its frequency of occurrence in textbooks. The predictive power of structural variables may derive in part from a correlation with frequency.

Campbell and Graham (1985; Graham, this volume) suggested that the order in which the arithmetic combinations are learned is another factor that may contribute to the predictive power of structural variables. Typically, within operations, the small-number combinations are encountered first and the larger problems are introduced later. Initially, there would be only a few candidate responses in the child's arithmecon, thereby providing few opportunities for associative confusions. As each combination is added to the structure, however, all prior existing responses constitute potential sources of interference (e.g. erroneous retrievals of these familiar responses establish false associations). Indeed, Norem and Knight (1930) found a correlation of 0.51 between difficulty rankings and the order in which the multiplication combinations were learned. Taken together, in the context of an associative interference account of accessibility, the effects of frequency of occurrence and order of acquisition may be sufficient to explain the problem-size effect. Norem's (1928) conclusion that difficulty is due largely to 'promiscuous connection forming' appears to be correct.

Implications for learning

While the evidence is strong that associative interference has a large impact on multiplication, Matthews (1985) has suggested that interference plays a smaller role in mental addition that in multiplication. Although he found evidence for associative interference, he also found several classes of problems (e.g. problems with 9 as an addend, problems summing to 10) that captured unique RT variance across the set of problems after variance associated with the interference variables (problem-error rate and the frequency of the correct sum as an error response) was removed. Matthews speculated that rule-based strategies (e.g. $9+X=(10+X)-1$) might be more common in addition than in multiplication. Such strategies are afforded in mental addition because the response set involves the positive integer number-line up to 18. Familiarity with this range of numbers may make counting, or analogue strategies (i.e. using a spatial image of the number-line to do addition) as reliable and effortless as retrieval in many specific cases. In contrast, fact retrieval may dominate multiplication (and perhaps mental division; Campbell 1985) because alternative strategies (e.g. counting) are relatively effortful and error-prone compared to memorization and retrieval.

When retrieval interference is a factor, however, the potential exists to make learning easier (cf. Graham, this volume). Minimizing interference means increasing the strength of correct associations relative to the strengths of false associations. To some extent, interference is an unavoidable consequence of the structure of arithmetic stimuli. For example, table-related errors in multiplication (e.g. $6 \times 9 = 48$) appear to reflect operand–product associations that are strong in the context of other problems (6 has a strong link to 48 via 6×8). If the development of associations was completely operand-driven, then each time a retrieval strengthened the association between the correct answer and the constituent operands of a problem, interference would be increased correspondingly for other problems involving those operands. In other words, to the extent that retrieval depends upon operand–product associations, one problem's gain is inevitably another problem's loss.

There may be a shift during learning, however, from operand-driven to problem-driven activation. When a new combination is first encountered, operand–answer associations from problems learned earlier may be strong relative to problem-driven activation, since the combination of operands is novel. Initially, erroneous retrievals may be promoted by operand–answer associations, but such retrievals would strengthen problem–answer links. With continued practice, problem-driven activation may come to dominate retrieval. Such a development could explain why Campbell and Graham (1985) found differences across operand order (e.g. 6×9 vs. 9×6) in adults, but no such effects in children. Furthermore, there is direct evidence that retrieval increasingly involves processing of the entire problem as a unit. There is a developmental trend for errors to be numerically close to the correct answer (Campbell and Graham 1985; Graham, this volume). This indicates that information based on the *combination* of operands in a problem is processed, since neither operand considered in isolation accurately predicts the magnitude of the correct product. It is the component of interference due to problem-driven activation that can potentially be controlled.

How might interference in arithmetic fact learning be minimized? It is now well established that problem difficulty (speed and errors) is positively correlated with how often a problem's correct answer occurs as an error response to other problems. This has been observed both in multiplication (Campbell and Graham 1985; Graham, this volume; Norem 1928) and in mental addition (Matthews 1985). In the interference theory, the empirical relationship between answer error rate, and both speed and accuracy of correct retrieval of an answer, means that a problem's susceptibility to forming interfering associations is higher when its correct answer becomes strongly associated with other problems. Conversely, the relationship can also mean that an answer is prone to entering into false associations when it has not formed a uniquely strong (preferrably correct) association. Two important predictions follow from a network-interference interpretation of the observed correlations between answer error rate and ease of correct retrieval of an answer. If strong correct associations are established for problems and answers encountered early in the learning sequence (i.e. a high criterion for speed and accuracy), those problems should be less susceptible to retroactive interference when other problems and answers are introduced later. Furthermore, the effects of proactive interference on later problems should also be reduced. Campbell and Graham (1985) showed that a substantial percentage of errors on multiplication problems encountered later in the learning sequence involved products to problems learned earlier. The correlations described above indicate that the intruding responses tend to be answers to problems that have not been learned well (i.e. are slow and error-prone). This suggests that if a high-performance learning criterion is set for problems introduced early on, then answers to these problems will have a lower probability of becoming sources of interference for later problems.

Although number-fact knowledge is but a small component in more complex numerical skills, it is a critical foundation upon which higher-level numerical problem-solving abilities depend (Resnick and Ford 1981). Indeed, the research by Siegler (this volume; see also Siegler and Shrager 1984) argues that the selection of arithmetic strategies by children depends directly upon the distribution of associations that determines retrieval performance. Consequently, understanding the role of interference in arithmetic fact learning may have far-reaching implications for the pedagogy of arithmetic. If difficulty is determined by 'promiscuous connection forming' as Norem (1928) suggested, then studying the structure of the problem sets and the structure of the learning conditions may reveal ways to minimize the impact of interfering associations and make learning the arithmetic facts substantially easier. Our children could still go forth and multiply, but there would be considerably less promiscuity.

Acknowledgements

Preparation of this paper was supported in part by a postgraduate scholarship from the Natural Sciences and Engineering Research Council of Canada (NSERC), and by NSERC grant A0790 to Dr. Neil Charness, Department of Psychology, University of Waterloo. Thanks are due to D. Besner, P. Brown, J. E. Cheeseman, N. Charness, D. Jeff Graham, D. Lavoie, and P. M. Merikle for helpful comments on an earlier draft.

References

Anderson, J. R. (1981). Interference: the relationship between response latency and response accuracy. *Journal of Experimental Psychology: Human Learning and Memory* **7**, 326–43.

Anderson, J. R. (1983). A spreading activation theory of memory. *Journal of Verbal Learning and Verbal Behavior* **22**, 261–95.

Ashcraft, M. H. (1982). The development of mental arithmetic: a chronometric approach. *Developmental Review* **2**, 213–36.

Ashcraft, M. H. (1983). Procedural knowledge versus fact retrieval in mental arithmetic: a reply to Baroody. *Developmental Review* **3**, 231–5.

Ashcraft, M. H. and Battaglia, J. (1978). Cognitive arithmetic: evidence for retrieval and decision processes in mental addition. *Journal of Experimental Psychology: Human Learning and Memory* **4**, 527–38.

Ashcraft, M. H. and Fierman, B. A. (1982). Mental addition in third, fourth, and sixth graders. *Journal of Experimental Child Psychology* **33**, 216–34.

Ashcraft, M. H. and Stazyk, E. H. (1981). Mental addition: a test of three verification models. *Memory and Cognition* **9**, 185–97.

Ashcraft, M. H., Fierman, B. A. and Bartolotta, R. (1984). The production and verification tasks in mental addition: an empirical comparison. *Developmental Review* **4**, 157–70.

Baroody, A. J. (1983). The development of procedural knowledge: an alternative explanation for chronometric trends of mental arithmetic. *Developmental Review* **3**, 225–30.

Campbell, J. I. D. (1984, May). *Go forth and multiply: the origins and dynamics of associative interference in mental computation.* Paper presented at the meeting of the Canadian Psychological Association, Ottawa, Ontario, Canada.

Campbell, J. I. D. (1985). *Associative interference in mental computation.* PhD. Thesis, University of Waterloo, Ontario, Canada.

Campbell, J. I. D. and Graham, D. J. (1985). Mental multiplication skill: structure, process, and acquisition. *Canadian Journal of Psychology* **39** (2), 338–66.

Clapp, F. L. (1924). The number combinations: their relative difficulty and the frequency of their appearance in text-books. *University of Wisconsin Bureau of Educational Research, Bulletin No. 2.*

Findlay, J. M. and Roberts, M. A. (March, 1985). *Knowledge of the answer and counting algorithms in simple arithmetic addition.* Paper presented at the Keele Conference on Maths and Maths Learning, University of Keele, Keele, England.

Groen, G. J. and Parkman, J. M. (1972). A chronometric analysis of simple addition. *Psychological Review* **79**, 329–43.

Matthews, S. (1985). *Associative Interference in Adult Mental Addition.* Unpublished B.A. Honours Thesis, University of Waterloo, Waterloo, Ontario, Canada.

Miller, K., Perlmutter, M., and Keating, D. (1984). Cognitive arithmetic: comparison of operations. *Journal of Experimental Psychology: Learning, Memory, and Cognition* **10**, 46–60.

Norem, G. M. (1928). *The learning of the one hundred multiplication combinations.* Unpublished M.A. Thesis, State University of Iowa, Arnes, IA.

Norem, G. M. and Knight, F. B. (1930). The learning of the 100 multiplication combinations. In *National Society for the Study of Education: Report on the Society's Committee on Arithmetic. Vol. 15, NSSE Yearbook 29*, pp. 551–67.

Parkman, J. M. (1972). Temporal aspects of simple multiplication and comparison. *Journal of Experimental Psychology* **95**, 437–44.

Parkman, J. M. and Groen, G. J. (1971). Temporal aspects of simple addition and comparison. *Journal of Experimental Psychology* **89**, 335–42.

Resnick, L. B. and Ford W. W. (1981). *The Psychology of Mathematics for Instruction.* Erlbaum, Hillsdale, NJ.

Siegler, R. S. and Shrager, J. (1984). Strategy choices in addition and subtraction: how do children know what to do? In *Origins of cognitive skills* (ed. C. Sophian) pp. 229–94. Lawrence Erlbaum Associates, Hillsdale, NJ.

Stazyk, E. H., Ashcraft, M. H., and Hamann, M. S. (1982). A network approach to simple multiplication. *Journal of Experimental Psychology: Learning, Memory, and Cognition* **8**, 320–35.

Suppes, P. and Groen, G. J. (1967). Some counting models of first grade performance data on simple addition facts. In *Research in mathematics education* (ed. J. M. Scandura). National Council of Teachers of Mathematics, Washington, DC.

Thorndike, E. L. (1922). *The psychology of arithmetic.* Macmillan, New York.

Wickelgren, W. A. (1976). Network strength theory of storage and retrieval dynamics. *Acta Psychologia* **41**, 67–85.

7

An associative retrieval model of arithmetic memory: how children learn to multiply

DAVID JEFFREY GRAHAM

The traditional teaching techniques for the simple multiplication facts may have made learning more difficult than necessary. Changes in response times and patterns of errors across grades are presented to support a model that emphasizes associative learning rather than computational learning. Predictions of the model were substantiated in an experiment in which teaching order was manipulated. The results are consistent with the view that a problem's relative difficulty is not determined by its size (the problem-size effect) but rather by the context in which it is learned.

Adults access an associative network of problems and answers when performing simple multiplication. The primary evidence for this view (and against procedural or computational accounts) is the very specific interference effects obtained in both arithmetic verification tasks (Stazyk, Ashcraft, and Hamann 1982) as well as in tasks in which numeric answers are generated (Campbell, this volume). In both tasks the retrieval of the correct response is inhibited by concurrent and/or prior activation of false responses that are related to the current stimulus in specific ways.

Campbell and Graham (1985) argue that these interfering associations are established in the initial stages of learning, when the child's network structure is first being formed. This paper begins with a review of some of the Campbell and Graham (1985) data showing the development of systematic errors in children's performance that reflect the functional dimensions of organization.

Developmental trends in errors and response latencies

Past research on simple multiplication has shown that the common errors made by people are highly systematic and very stable across samples; (Campbell 1984; Campbell and Graham 1985; Norem 1928). Three mutually exclusive classes of errors can be identified. The first is *table-related,* which are errors that are answers to other problems in one of the times-tables defined by a given problem's operands (e.g. $4 \times 5 = 16$, and $6 \times 7 = 49$). *Table-unrelated* errors are answers to problems in unrelated times-tables (e.g. $4 \times 5 = 18$, and $6 \times 7 = 45$). Finally, a third catch-all class is *miscellaneous,* errors that are not answers to any simple multiplication problem (e.g. $4 \times 5 = 17$, and $6 \times 7 = 47$).

If errors are made by guessing randomly, then approximately 14 per cent would be table-related, 19 per cent would be table-unrelated, and 67 per cent would be miscellaneous (see Campbell and Graham 1985). The data reported by Norem (1928) and Campbell (1984) indicate that table-related errors occur much more often than could be expected by chance (over 70 per cent vs. 14 per cent), indicating that errors are systematic rather than random. By examining the nature of error production, the principles that govern retrieval on correct trials can be inferred and experimentally tested as Campbell (this volume) does with error-priming.

Campbell and Graham (1985) collected a large body of errors and response times on simple multiplication problems in order to examine developmental trends in the types of errors children make and the factors that govern how fast they can respond correctly (for a detailed description of the method, procedure, and summary results, see Campbell and Graham (1985)). It is generally true that problems were taught in the context of times-tables and practised in both operand orders (e.g. 5×8 and 8×5). The grade-3, -4, and -5 classes were tested in March 1984. The grade-3 class was tested twice again, at approximately six-week intervals, and the grade-2 class was tested in June, after one month of training with problems up to the 5 times-table.

The problem set

All problems between 0×0 and 9×9 were tested once in each operand order. Tie problems (e.g. 7×7) were tested twice. The grade-2 class was tested only on problems up to 6×6, omitting the other problems with both operands greater than or equal to six. For the purposes of this paper, data are reported only for the 36 problems from 2×2 to 9×9. The 0 and 1 times-tables (which tend to be faster, on average, than the 2 times-table) do not require retrieval in the same sense as the other problems, since they can be solved by the fast application of procedural rules (e.g. $N \times 0 = 0$, or $N \times 1 = N$).

Apparatus and procedure

Children were tested one at a time by means of a microcomputer. The 45 non-tie problems and the 10 tie problems were presented in each of two blocks, where operand order was randomly determined in block one and reversed in block two. Children were instructed to respond as quickly and accurately as they could, and response times and any errors were recorded on each trial. The experimenter watched carefully for finger counting or lip movements that could indicate internal counting and noted the number of trials in which they were present.

Results and discussion

Table 7.1 contains error data for grades 3, 4, and 5 and adults with the chance estimates discussed above (Campbell and Graham 1985). The numbers reported in the top of Table 7.1 represent the 'relatedness' category effects, showing the proportion of total errors made that fall in each of the error classes for the problems from 2 × 2 to 9 × 9. The three sessions for the grade-3 class are shown in the left-hand panel while the trends across grades

Table 7.1. *(a) Per cent of errors that were table-related (TR), table-unrelated (TU), or miscellaneous (MI) across grade-3 sessions and groups, and (b) per cent of errors within each relatedness class that were within one maximum operand unit of the correct product.*

Error class	Session[a]				Group[b]			
	3A	3B	3C	Chance	3A	4	5	Adult
(a) *Relatedness category effects*								
TR	43.2	47.7	56.2	14.2	43.2	47.6	68.5	79.1
TU	20.8	21.0	21.4	19.1	20.8	21.5	14.3	13.5
MI	36.0	31.3	22.4	66.7	36.0	30.9	17.2	7.4
(b) *Magnitude category effects*								
TR					54.7	68.4	74.4	25.5
TU					36.3	49.1	62.2	13.1
MI					53.7	61.1	64.8	11.4

[a] 3A, 3B, and 3C correspond to the first, second, and third testing sessions for grade 3, respectively.

[b] 3A = grade 3, first session; 4 = grade 4; 5 = grade 5. Note: total number of errors per group: 3A = 491, 4 = 492, 5 = 314, adult = 661. *N* for each group: 3A = 23, 4 = 27, 5 = 26, adult = 60.

are shown to the right. Across both sessions and grades, table-related errors began to predominate. Conversely, miscellaneous errors started out low and dropped continually to only 7.4 per cent in adults. Table-unrelated errors started out around chance levels for grades 3 and 4, and dropped by grade-5 to adult levels of performance.

Errors tended to be systematic even in the youngest age-groups. In the first session of grade 3, table-related errors were about 30 per cent above chance and miscellaneous errors were 30 per cent below chance. Moreover, the grade-2 data show that even after the first month of training, 46 per cent of the errors made were table-related, while only 25 per cent would be expected by chance with the 27 problems from 2×2 to 6×6. Why is it that smaller and smaller subsets of possible integers are consistently promoted as errors? These data suggest that after the first introduction to the multiplication problems and answers, children very quickly become familiar with a set of responses that are multiplication answers, although they are not certain which problems they belong to.

As more experience is gained, most of the errors are correct answers to problems in related times-tables. It appears as if the retrieval system begins to discriminate between merely familiar items and those that are more likely correct. One way in which this could occur is by a type of category-priming. Each of the operands in the problem serves as a label for a times-table that has associated with it a set of answers. An operand activates all of its respective times-table's answers, using associative links that have developed over a large number of correct trials in which the operand has been paired with eight different products. Errors seem to reflect intrusions from associations with an operand that is correct in the context of another problem. For example, the response 24 to the problem 4×5 may be promoted by 4's association with 24 in the problem 4×6. The strength of such associations would be primarily frequency based, implying that incorrect retrievals would make themselves more likely to occur again later, since that error having been made in the first place establishes a link between that false product and the given problem.

This account explains errors as the result of the nature of the associative network, with links being established between operands and candidate responses. However, retrieval cannot be all operand-driven. There is evidence that a problem may also be treated as a unit, and that it has associated with it a rough magnitude representation. Across grades, and across grade-3 sessions, there was a clear trend for error responses to converge around the correct product. In order to quantify this phenomenon a measure was devised that classified errors according to how far off they were from the correct product. If the absolute difference between the error and the correct product was less than or equal to the size of the maximum operand in the problem, the error is called a *near* error; for example, the response 24 (correct to 4×6) would be within one maximum operand unit of the correct answer for 5×6.

Convergence around the correct product is observed when a higher percentage of total errors falls in the near class, as subjects gain more experience. This can be seen in the bottom panel of Table 7.1 representing the 'magnitude' category effect. In each of the relatedness categories, the proportion of near errors increased across grades. Errors that are far off the correct product occur less often, as children learn. Campbell and Graham (1985) interpreted this trend as evidence for problem-driven activation. Associated with the problem as a whole is an abstract magnitude; that is, an estimate of the range in which the answer is likely to fall. How this would affect retrieval was not specified.

The concept of feature-overlap (e.g. Flexser and Tulving 1978) may be a relevant mechanism to account for this type of associative behaviour. The problem node will have a set of defining features, part of which are those which represent magnitude information (whether this is analogue or digital is still controversial). The set of response candidates activated by the encoding of particular operands in a problem will be further activated to the degree that they have highly similar magnitude representations with the problems node. In this way, candidates that are associated by means of times-tables may not have magnitude attributes in common with or overlapping with those defined by encoding the problem. Thus, during processing they would be rejected early as viable selections (e.g. the answer 12 is a table-related error to the problem 6×5 but would not likely occur since the magnitudes associated with the problem and that answer are incongruent).

Other evidence for a magnitude estimation process comes from verification data (Stazyk *et al.* 1982). When false products are very far off the correct product (i.e. large split), those trials are very easy to reject. In fact, they can be rejected faster than their corresponding correct trials, suggesting that the magnitude associated with the problem is available fast enough to reject an implausible candidate, without requiring an explicit retrieval and comparison with the correct product. Stazyk *et al.* argue that a magnitude estimation process goes on in parallel with retrieval processes, thereby accounting for the fast rejections at very large splits. In an associative model, magnitude is a primitive featural representation common to all nodes in the numeric network. It is the unique semantic component of number knowledge. Very fast congruency checks between the magnitudes associated with the given problem and foil are possible as soon as they are encoded, creating the RT advantages observed on large-split trails. The estimation process is applied to the problem stimulus and need not depend on the retrieval of the correct response.

The developmental trends for the predominance of table-related errors (the relatedness effect) and the convergence around the correct product (the magnitude effect) define the basic associative dimensions around which the numeric memory system is organized for multiplication. The underlying assumption is that errors reflect associative relations between problems and answers (Siegler, this volume; Siegler and Shrager 1984), an approach not unlike Conrad's (1964) use of acoustic confusion errors to infer phonological coding in immediate memory. As the numeric network is being established through practice, the major associative pathways are (1) operand-driven activation of times-table answer sets, and (2) problem-driven activation of a set of answers that are congruent in magnitude with the problem's magnitude attributes. If these types of activation sum together then table-related errors that are close to the correct answer will become the predominant type of error.

The network-interference model of arithmetic retrieval

In a detailed description of the numeric memory system there will need to be ways to implement the use of computational rules, the distinction between numbers as operands and numbers as products, and the representation of a problem as a unit. To start with, however, an associative model is presented here to incorporate the observed effects of familiarity, table-relatedness, and magnitude similarity, and the principled relation between errors and retrieval time (Campbell, this volume; Campbell and Graham 1985). If retrieval is accomplished by spreading activation, then, in the future, we should be able to find evidence for both priming and interference, just as in lexical decision and sentence verification tasks. For now, the model is developed to handle interference effects.

The first associative dimension to implement is times-table category membershp. In the bottom of Table 7.2 the response candidates are shown for a particular trial on the problem 6×9. Each candidate has a probability of being selected that is reflected in the error frequency counts shown directly above. The sequential encoding of each of the operands serves to activate two sets of responses in the times-tables specified by the operands. The sets activated by each operand are given in the two rows indicated by $6\times$ and $9\times$. Candidates activated by both operands are plotted in the row indicated by 6,9, while other common responses activated by the task context are presented in the row indicated by $\times$. In the top panel, frequency

Table 7.2. *Product frequency counts (errors per 1000 trials for 6×9) showing active candidate sets (in bold) across age-groups.*

Sample	Products																
	18	20	24	27	28	30	32	36	40	42	45	48	54	56	63	72	81
Adults	0	0	0	0	0	**8**	**4**	**58**	**0**	**17**	**17**	**8**	**cp**[a]	**79**	**50**	8	5
Children	7	0	**7**	**7**	**7**	**0**	**13**	**47**	**0**	**27**	**33**	**13**	**cp**[a]	**80**	**40**	**20**	**0**
Response candidates																	
×[b]		20			28		32		40					56			
6×[c] 6			24			30				42		48					
6, 9[d]	18							36					54				
9×[e] 9				27							45				63	72	81

[a] Correct product.
[b] Familiar answers in the context of multiplication.
[c] Answers to 6 times-table.
[d] Answers in both 6 and 9 times-tables.
[e] Answers to 9 times-table.

counts (expressed as the number of times that answer could be expected out of 1000 trials) are given for each response from the adult sample and pooled from grades 3A, 4, and 5 (Campbell and Graham 1985).

Operands act as category labels, just as saying 'animals' or 'colours' will activate instances of those classes in semantic memory. One or more of these instances are sufficiently activated to enter conscious awareness or attract attentional resources, though most are not. In an established arithmetic network the operand 6 is associated with the multiples of 6 from 6 to 54, and the 9 activates the multiples from 9 to 81. There are three numbers (18, 36, and 54) that receive activation from both of the operands (due to correct digit-product traces from other problems; e.g. 6×6 and 9×4). This double activation makes them more likely to be selected than most of the other candidates (36 is a very common error for 6×9, occuring about twice as often as 42 or 45). Even though 18 is activated by both operands, it is rarely given as an error, presumably because the magnitude associated with it is too small to be an answer to 6×9. The answer 36 is within a plausible range.

The abstract magnitude associated with the problem as a whole serves to boost those candidates whose magnitudes are within a plausible range of values. The amount of activation received by a candidate is proportional to the degree of its correspondence with the problem's magnitude attributes. In the early grades when the problem's magnitude is not well defined, a subset of responses (shown in bold in the children's line in Table 7.2) receives activation in addition to that caused by category membership by virtue of its common magnitude shared with the problem. As more experience is gained with the problem, the number of highly active responses reduces to a smaller subset of answers (shown in bold in the adult's line in Table 7.2). (It may be that each operand combination develops a well-defined magnitude simply by being repeatedly paired with a product that already has a well-defined magnitude.)

The correct product receives a triple boost of activation making it the most likely response. However, under speeded instructions, subjects may select the first candidate that surpasses some threshold level of activation. Each of the responses in the working candidate set will vary in its associative strength with the given problem, so that numbers that are closer to the correct product will have higher associative strengths. This is consistent with the magnitude effect observed in the children's data, in which the probability of a given error decreases as distance from the correct product increases.

In this model, correct retrieval time (RT) is determined, in part, by the total strength of false associates activated by the problem (Anderson 1983). If a particular candidate is selected when its activation level exceeds a critical proportion of total activation, then, the greater the amount of false activation in the system, the longer it will take to elicit the correct response. In other words, when a problem is encoded, a number of response candidates are activated according to their relationships with each operand and with the stimulus as a whole. The candidates that are the most highly activated compete for selection. While the exact nature of this competition process remains unspecified, its impact on RT will vary directly with the size of the activated response set and its relative associative strength. We can derive an estimate of the total strength of false associates using an assumption underlying Siegler and Shrager's (1984) concept of the distribution of associations: the probability of selecting any particular response is directly proportional to its associative strength. Over a sufficient number of observations on a problem, a variety of errors will occur that vary in their relative frequencies (and thus in their relative associative strengths). The total number of errors on a problem can therefore be used as an estimate of the amount of false activation in the candidate set.

If the hypothesis that candidate competition influences retrieval of the correct product is true, then we should expect to find a strong positive correlation between correct RT and the *total error* rates (TE) across the problem set (and not a speed/accuracy trade-off). The dynamics of the retrieval system are such that correct RT for a problem will be long if there are many competing candidates activated. In addition, the larger the active candidate set the higher the overall probability of an error.

Competition among candidates is not the only source of interference. A second potential source, hinted at by Norem (1982), and developed in Campbell and Graham (1985), is false associations with the correct product. The frequency with which a problem's correct product occurs as an error on other problems is called a *product error* rate (PE). False associations with a problem's correct product may have effects that are independent of those caused by false associations generated by the problem. Two distinct associative processes may be involved; candidate competition for a limited amount of activation at the correct response node, in which associates to that response siphon off activation as it accumulates (cf. Anderson 1983). The greater the number of connections at that response node, the longer it will take to accumulate the required amount of activation. If this dual hypothesis is correct then problems' total error rates (TE) and product error rates (PE) should both contribute, independently, to predicting correct solution times. These error measures will be referred to as interference variables.

Explaining differential problem difficulty

A network-interference model offers an alternative explanation for the problem-size effect (see Ashcraft 1982; Baroody 1983; Parkman 1972). Correlations as high as 0.60 to 0.80 have been reported between RT and a variety of problem-size measures called *structural variables*. These include the numeric magnitude of the minimum operand, the maximum operand, the sum of the operands, and the correct product. Structural variables are interpreted as indices of the answers' location in the network (e.g. Ashcraft 1982) or the number of operations involved in automatized counting procedures (or procedural complexity; see Baroody 1983). In these accounts, larger problems involve longer response times either because of the distance searched in memory or the number of operations required to compute the answers.

The account proposed in this paper is radically different. Structural predictors do not tell us anything directly about the nature of the operations involved. They are only coincidentally related to problem RTs, confounded with factors that influence the dynamics of retrieval, namely, differences across problems in testing frequency, and the temporal order in which problems are usually taught. Campbell and Graham (1985) wanted to determine whether or not differences in correct RT across problems could be explained in terms of differential susceptibilty to interference effects. In order to test this notion, multiple regression analyses were designed to compare the amount of variance captured by the structural variables with that captured by interference variables. Do structural variables contribute to predicting correct RT, once variance due to interference is accounted for? If they do then an interference account would be insufficient. Secondly, the model proposes two separate sources of interference, and therefore the two error measures (TE and PE) should contribute independently to the regression equations.

Design of the regression analyses

The analyses reported in this paper employed eight variables to predict correct RTs across the 36 problems (27 for grade 2). The predictors included the five common structural variables; minimum operand (MIN), maximum operand (MAX), sum of operands (SUM), SUM squared (SSQ), and correct product (CP). There were the two interference variables: total problem error rates (TE), and product error rates (PE). The last predictor was a tie/non-tie vector (TIE) coding tie problems as 1 and non-ties as 0. This is used in order to test the idea that the ties' advantage is due to there being only one unique operand and thus smaller candidate sets. If this is true then once variance due to interference is factored out (assuming interference is a function of candidate set size) there should be no differences between ties and non-ties. Regression analyses were performed for each class in order to observe any developmental trends in the relative importance of structural and interference variables. The simple correlations used in these regressions are presented in Table 7.3.

There are a number of systematic patterns that need to be considered. While TE yields the highest correlation with RT at all grade levels, there is

Table 7.3. *Correlations used in the multiple regression analyses predicting correct retrieval time with structural and interference variables (see text for definitions of predictors). Note: correlations are based on 36 pairs (except grade 2 = 27). Critical r's = r(25) = 0.38, r(34) = 0.33.*

Grade	Interference set		Structural set					
	TE	PE	MIN	MAX	SUM	SSQ	CP	TIE
2	0.77	0.52	0.63	0.60	0.76	0.74	0.76	−0.24
3A	0.75	0.49	0.53	0.69	0.70	0.68	0.66	−0.36
3B	0.81	0.57	0.52	0.63	0.66	0.63	0.61	−0.32
3C	0.80	0.77	0.41	0.63	0.60	0.55	0.52	−0.37
4	0.69	0.59	0.24	0.56	0.47	0.39	0.35	−0.45
5	0.76	0.73	0.27	0.62	0.51	0.43	0.38	−0.41
Adult	0.85	0.45	0.45	0.72	0.68	0.64	0.60	−0.40

no consistent leader amongst the structural variables. SUM, SSQ, and CP show equally large coefficients in the early grades, and there is a trend for MAX to be best in later grades (although not significantly). Notice that all structural variables (except MAX) show a systematic decrease in size across grades, suggesting that their importance diminishes. Conversely, the TIE vector gradually increases across grades, as does the product error (PE) correlation.

The major difficulty in interpreting these simple correlations is that all of the predictors are highly intercorrelated, and only 3 of the 56 coefficients are non-significant. A clearer picture is available from the multiple regressions in which the common variance is partialled out, allowing us to consider the unique variance in RT that each of the predictors can account for.

Regression results and discussion

Entering the interference variables first

In keeping with the theoretical importance of the interference variables, Campbell and Graham (1985) entered these predictors first in the regression equation. Following these, the tie/non-tie vector was entered to test the competition account of their special status. Only those variables that contributed significantly remained in the equation and then structural variables were allowed to enter. The rationale for this entry procedure was to determine whether or not the interference factors are sufficient to account for differential problem RTs. If structural variables are of importance in their own right, then they should be able to capture a significant portion of the remaining variance.

Problem errors (TE) accounted for about 57 per cent of the variance on average. Confirming the presence of separate sources of interference, product errors (PE) always contributed independently to the regression equations, capturing on average an additional 10 per cent of the variance. The only differences between grades occur with the third variable to enter significantly. Contrary to the prediction based on set sizes, the TIE vector captured an additional 7.6 per cent of the variance in all grades, except grade 2. In grade 2 the correct product (CP) entered third, with about 6 per cent of the variance. In the adult data, SUM entered fourth, after TIEs, with 3 per cent of the variance.

Structural variables failed to capture any unique variance in five out of seven samples, and less then 5 per cent, on average, in the grade 2 and adult samples. Campbell and Graham (1985) argued that, given these failures, the standard interpretations of structural variable correlations as indices of processes that govern correct RT are severely weakened. While their entry procedure was theoretically motivated, it might be argued that it is not a convincing test of the relative validity of structural vs. interference models. The amount of variance captured by a particular variable depends upon the step upon which it enters the regression equation. By entering the interference variables first, followed by TIE, we minimized the likelihood that any structural variable could enter at all. A more convincing test would be to use a stepwise free-entry procedure, allowing all variables a chance to enter at each step. If a variable's final beta weight is significantly different from zero, then it captures unique variance in problem RTs and must be accounted for in the model. For the purposes of this paper the analyses were rerun with appropriate modifications.

Entering all predictors stepwise

Table 7.4 contains the results of the stepwise free-entry procedure using all predictors in one block. There were few changes in the final regression equations when this procedure was employed. Since TE always had the highest simple correlation, it is no surprise that it consistently enters first. The important point to draw is that, once again, no structural variable enters before PE and TIE, in five of the seven samples. Compared to the earlier analyses, the equations for grades 2 and 5 remained the same, and the grade-3 and -4 data found TIE entering second and PE entering third. PE did not enter significantly in grade 3B or adults, although tests of its beta weights after the last step yielded p values of 0.14 and 0.06 respectively. Instead SUM entered third for grade 3B, and MAX entered second for adults. Overall, the same pattern occurred as in the initial analyses, with the interference variables capturing about 66 per cent of the total variance.

Entering the structural variables first

The fact that structural variables account for only 5 per cent of variance in only some of the groups is not necessarily damaging to models that rely on structural variable correlations with RT. One could argue that the structural

Table 7.4. *Group summaries of stepwise free-entry multiple regressions.*

Predictor[a] [b]		Group						
		2	3A	3B	3C	4	5	Adult
	STEP	1	1	1	1	1	1	1
TE	VAR	0.59	0.57	0.66	0.63	0.39	0.57	0.77
	BETA	0.393	0.711	0.639	0.600	0.464	0.543	0.643
	STEP	2	3		3	3	2	
PE	VAR	0.08	0.03	—	0.06	0.10	0.14	—
	BETA	0.269	0.194		0.328	0.353	0.342	
	STEP							2
MAX	VAR	—	—	—	—	—	—	0.05
	BETA							0.261
	STEP			3				
SUM	VAR	—	—	0.04	—	—	—	—
	BETA			0.262				
	STEP	3						
CP	VAR	0.06	—	—	—	—	—	—
	BETA	0.376						
	STEP		2	2	2	2	3	3
TIE	VAR	—	0.16	0.09	0.16	0.16	0.04	0.03
	BETA		−0.370	−0.302	−0.304	−0.343	−0.223	0.191

[a] See text for definitions.

[b] Step: order of entry in equation; var: variance accounted for on entry; beta: final standardized beta weight.

properties of problems predict error rates as well as RT. For example, an automatic counting model could assume that error rates would increase with the number of incrementing steps required to compute the answer, as a function of the size of the operands. (Such a model would have difficulty explaining product error correlations, since errors on one counting sequence are somehow governing correct solution times on completely separate counting sequences.) However, given that these interference correlations could be explained, when we factor out TE and PE in the first steps, a large proportion of the variance that would be accounted for by structural variables is removed. This type of argument would predict that, if we were to let structural variables enter first, there should be very little, if any, remaining variance to be accounted for by the interference factors.

To test this prediction, the structural variables were entered stepwise in the first block and the interference variables were entered stepwise in the second block. Only one of the structural variables entered in each class, accounting for about 45 per cent of total variance on average. When the second block of variables was entered, the beta weights for the structural variables dropped to zero (i.e. were non-significant) in all groups except grade 2 and adults. Interference measures still contributed independently once variance associated with the structurals was removed, accounting for an additional 26 per cent of the variance, with both TE and PE contributing independently in all groups except adults.

Conclusions from the regression analyses

These analyses support the conclusion that an interference account is sufficient to explain differences in problem difficulty (assuming that error rates index the influence of false associations). This is strengthened when one considers the relative amounts of unique variance explained by the two classes of predictors. Secondly, the grade-2 data indicate that the effects of false associations on response time are present after only one month of training (although evidence of intentional counting strategies on about 16 per cent of trials was noted which have additional effects on RT). Thirdly, given that the TE and PE measures account for unique variance, support was obtained for two sources of interference: correct retrievals are inhibited both by false associations with the problem and by false associations with the product.

One problem remains to be explained if one is to accept the interference account of differential problem difficulty. Why should bigger problems be more susceptible to interference than smaller problems? In other words, why do structural variables predict RT? The next section of this paper addresses this issue by arguing that structural variables are confounded with practice frequency and order of acquisition, factors that affect trace formation in the retrieval structure.

Effects of teaching order on problem difficulty

The standard teaching order for simple multiplication problems starts with small problems and proceeds through the times-tables from the twos through the nines. If problems learned earlier are reviewed when new problems are encountered then small problems will receive more practice initially than larger problems. Obviously, after all problems have been practised for long enough, the initial differences should become marginal. However, in a frequency count of simple problems in mathematics textbooks (Clapp 1924; Thorndike 1922), smaller problems are tested more often than larger problems. If this is true for today's textbooks the initial practice differences may be maintained over a number of years of schooling.

Problems learned earlier will benefit in a second manner as well. There will be a smaller subset of problems and answers established in the network when they are encountered for the first time. These problems will have fewer competing associations to hinder learning, promoting stronger correct associations which are less susceptible to interference. Conversely, problems learned later in the sequence will be taught in the context of a larger set of competing responses, thereby producing weaker correct associations that are more susceptible to interference. Furthermore, individual operands in the later problems will encounter more proactive interference from established associations. This hypothesis was tested in the following experiment in which two groups of grade-3 children were compared under different teaching orders.

Teaching order and the problem-size effect

Given the cumulative interference hypothesis, one should be able to reverse the problem-size effect by teaching problems starting at 9×9 and going smaller. Then, whatever index is used to measure problem size should be negatively correlated with RT. Such an extreme manipulation could have unforeseeable adverse effects on children learning multiplication, so a more conservative approach was adopted. One group was taught in the standard order and the other group was taught in a mixed order.

Method and procedure

The 1984–1985 grade-3 class was split into two groups, matched for general mathematics ability by the teacher. Both groups were taught the 2 times-table first, in order to have a baseline measure and to familiarize students with concepts and the teaching procedures. Then the next 18 problems (from 3×3 to 5×9) were divided into 3 sets of problems which would be practised one at a time for five days before moving on to the next problem set. The teaching method was the same for both groups. Although an emphasis was placed on rote memory exercises such as flash-cards and drill-sheets, the daily regimen included more conceptual drills such as the grouping of concrete objects into rows and columns to represent given problems. All problems were tested equally often, yet some variance due to practice at home was unavoidable.

The *table-order* group ($N = 10$) was given problem sets that corresponded roughly to times-tables of increasing size. For the *mixed-order* group ($N = 10$), problem sets 2 to 4 were composed of the same 18 problems as in the corresponding table-order sets, only problems were counterbalanced so that the sets would be roughly equal in the mean minimum and maximum operands and in mean products. At the time of first testing the problems from 6×6 to 9×9 had not yet been introduced to either group and results for these problems are not reported. Error responses and correct response times were obtained from subjects tested individually by means of a PET 2001 microcomputer. Each of the 26 problems was presented twice in each operand order.

Results and discussion

The predicted effect on the structural variable by RT correlations was obtained. These correlations were calculated on the 26 problems from 2×2 to 5×9, with group means calculated for each problem collapsing over operand order. The best structural predictor for the table-order group was correct product ($r = 0.84$), while the sum was the best for the mixed-order group ($r = 0.68$). While this difference is statistically marginal, the fact that

Table 7.5. *Simple correlations and regression analyses with interference and structural predictors of retrieval time for each learning-order group (table-order and mixed-order).*

Predictor[a]	Analyses						
	Correlations		[b]	Stepwise		Structural first	
	Table	Mixed		Table	Mixed	Table	Mixed
			STEP	1	1	2	2
TE	0.90	0.79	VAR	0.81	0.62	0.14	0.19
			BETA	0.645	0.786	0.447	0.607
			STEP			3	
PE	0.72	0.70	VAR	—	—	0.03	—
			BETA			0.226	
			STEP	2			1[c]
SUM	0.83	0.68	VAR	0.06	—	—	0.46
			BETA	0.340			0.253
			STEP			1	
CP	0.84	0.64	VAR	—	—	0.71	—
			BETA			0.363	
			STEP				
TIE	−0.43	−0.47	VAR	—	—	—	—
			BETA				

[a] See text for definitions.
[b] Step: order of entry in equation; var: variance accounted for on entry; beta: final standardized beta weight.
[c] Non-significant final beta; $p > 0.05$.

all structural correlations are consistently lower for the mixed group is suggestive. When the order of learning is changed the amount of variance accounted for by structural variables drops. The summary of simple correlations for each group for structural and interference predictors is given in Table 7.5. Notice that the problem errors (TE) give the highest correlation in both groups. Product errors (PE) yield significant correlations of the same size for both groups.

Two multiple regressions were performed for both groups: (1) entering all variables stepwise, and (2) entering structural variables first (see Table 7.5). In these samples, PE entered only once, in the table-order group, after the correct product (CP) was entered. However, in line with the predicted teaching-order effect, both TE and either SUM or CP entered the equations for the table group, while no structural predictors remained significant for the mixed group. These results must be interpreted cautiously due to the small samples employed; however, the fact that structural variables did not contribute independently in any mixed-group analysis lends credence to the view that the predictive ability of structural variables is largely determined by teaching order.

An analysis of variance was performed to cross-check the correlational data. A learning order (table and mixed) by problem-set (3-, 4- and 5-times-table problems) design was used incorporating the 18 problems from 3×3 to 5×9. For both groups the problem sets were defined by the sets given to the table-order group; for example, set 1 consisted of $3 \times 3, 3 \times 4, \ldots, 3 \times 8$. Even though the mixed group would have learned these 6 problems in three different sets, they would now be classified as set-1 data. If teaching order has no effect on the relative difficulty of problems then the groups should show the same rank-order RTs across the problem sets.

There was no effect of group ($F < 1.0$) or of set ($F(2,18) = 1.8, p > 0.05$); however, there was a significant interaction between learning order (group) and set ($F(2,36) = 3.4, p < 0.05$). Tests of the simple main effects revealed that there was no effect of set for the mixed group ($F(2,36) = 1.01, p > 0.05$) and a large set effect for the table group ($F(2,36) = 4.2$, $p < 0.05$). The analysis of errors revealed no effects of group nor any interaction, but there was a main effect of set, in which the 3-times-table set was much less error-prone than the 4- and 5-times-table sets for each group.

This study provides evidence that changing the order in which problems are taught reduces the size of the correlations with structural variables and changes the relative speed with which problems are answered. If these effects are replicated in larger samples they would be damaging to any account that emphasizes the functional relevance of structural variables. Thus, larger problems require longer RTs, not because of the number of operations needed, nor because of their location in a sequentially scanned numeric network; they are slower because they are more susceptible to interference factors, by virtue of the order in which they are learned, and the number of competing associations that make up the learning context.

Educational implications

The evidence to date suggests that multiplication performance is governed by a retrieval process from an associative network of problems and answers.

The network's structure is organized on the basis of times-table membership and shared magnitude attributes. The specific structure that evolves in early learning is sensitive to such factors as familiarity with the response set, the frequency with which particular problems are practised, and, perhaps most importantly, the order in which problems are learned. Based on these theoretical notions one may be able to develop an optimal teaching order and practice regimen which will reduce the amount of class-room time needed for simple multiplication. Over the long run, more time can be spent on complex problem-solving, and retrieval errors could be minimized.

A training schedule for simple multiplication facts should emphasize accuracy, as long as response times are reasonably fast. In order to promote accuracy, associative confusions due to times-table relations could be minimized by not teaching problems in the context of times-tables. Perhaps this long-time tradition has mislead educators to believe that a well-ordered systematic introduction to the facts provides beginners with a helpful conceptual framework. It is still very much an empirical question as to whether alternative orders are better. One such order could be to divide the 36 problems (2×3 to 9×9) into 6 sets of 6 problems. Each set would contain problems from many times-tables by using as many different operands as possible. These would be constructed so that the products in each set are widely divergent in magnitude, and are rarely common errors to other problems in the same set (based on normative data). Furthermore, it may prove beneficial to give the more difficult problems (e.g. 3×8, 4×7, 6×9, and 6×7) a head-start. This could be done by placing them in the first set and requiring a strict performance criterion before moving on to new sets.

There is a subset of problems whose products (12, 16, 18, 24, and 36) are particularly prone to error because they are correct to more than one problem. It may be better to teach these 10 facts after the other problems have been mastered. In this way these products would be less likely to form false pairings in the early stages of learning. As well as order considerations, any explicit heuristics that could limit the size of a problem's candidate set should speed up acquisition time. For example, the 5 times-table may be easier because all of its products end in a 0 or a 5. Likewise, the 9 times-table has a pattern it follows in which the digits in the response add to nine. Making use of incidental relations between operands and responses can aid on a number of isolated problems (e.g. $56 = 7 \times 8$ matches a counting string).

Computerized drills that deal with individualized needs are quite feasible with the advent of computers in the class-room. Review programs could be designed such that all problems are periodically maintained while error-prone problems and responses are concentrated on. For example, if a problem like 7×8 has a recent history of errors composed primarily of the responses 54 or 48, then it would be tested more often in conjunction with the problems 6 × 9 and 6 × 8. Computer drills would be much more practical than paper-and-pencil tests since more trials could be run and error histories automatically recorded. Moreover, instant feedback at a computer screen would help bolster the correct associations and minimize the impact of false retrievals.

Acknowledgements

This research was supported by a postgraduate scholarship to the author from the Natural Sciences and Engineering Research Council of Canada, and in part by NSERC grant 037–6458 awarded to Dr Neil Charness. I am grateful to D. W. Higgins, and K. Thompson of the Perth County Board of Education, as well as to Janice Graham and Dawn Franklin for their practical advice and co-operation. Thanks go to Jamie Campbell for his interest and collaboration throughout all phases of the research project.
Reprint requests should be sent to Jeff Graham, Department of Psychology, University of Waterloo, Waterloo, Ontario N2L 3G1, Canada.

References

Anderson, J. R. (1983). A spreading activation theory of memory. *Journal of Verbal Learning and Verbal Behaviour* **22,** 261–95.

Ashcraft, M. H. (1982). The development of mental arithmetic: a chronometric approach. *Developmental Review* **3,** 231–5.

Baroody, A. J. (1983). The development of procedural knowledge: an alternative explanation for chronometric trends of mental arithmetic. *Developmental Review* **3,** 225–30.

Campbell J. I. D. (1984, May). *Go forth and multiply: the origins and dynamics of associative interference in mental computation.* Paper presented at the meeting of the Canadian Psychological Association, Ottawa, Ontario, Canada.

Campbell, J. I. D. (1985). *Associative interference in mental computation.* PhD. Thesis, University of Waterloo, Waterloo, Ontario, Canada.

Campbell, J. I. D. and Graham, D. J. (1985). Mental multiplication skill: structure, process, and acquisition. *Canadian Journal of Psychology* **39** (2), 338–66.

Clapp, F. L. (1924). *The number combinations: their relative difficulty and the frequency of their appearance in textbooks.* Bulletin No. 2, University of Wisconsin Bureau of Educational Research.

Conrad, R. (1964). Acoustic confusions in immediate memory. *British Journal of Psychology* **55,** 75–84.

Flexser, A. J. and Tulving, E. (1978). Retrieval independence in recognition and recall. *Psychological Review* **85,** 153–71.

Norem, G. M. (1982). *The learning of the one hundred multiplication combinations.* Unpublished MA Thesis, State University of Iowa, Ames, IA.

Parkman, J. M. (1972). Temporal aspects of simple multiplication and comparison. *Journal of Experimental Psychology* **95,** 437–44.

Siegler, R. S. and Shrager, J. (1984). A model of strategy choice. In *The origins of cognitive skills* (ed. C. Sophian). Erlbaum, Hillsdale, NJ.

Stazyk, E. H., Ashcraft, M. H. and Hamann, M. S. (1982). A network approach to simple multiplication. *Journal of Experimental Psychology: Learning, Memory, and Cognition* **8,** 320–35.

Thorndike, E. L. (1922). *The psychology of arithmetic.* Macmillan, New York.

8

The internal representation of number: analogue or digital?

R. R. TODD, P. J. BARBER, and D. JONES

Introduction

This paper contrasts two possible forms of internal representation in one of the most basic of mathematical concepts, that of number. In particular, it focuses on the construction of analogue representations of number, as distinct from digital representations. These twin facets of number as represented have different roots in cultural evolution (Wilden 1972) and different functions (Wilder 1968). One study is reported which attempts to explore the form of the analogue, and to observe developmental changes during the school years, and a second study is described that attempts to evaluate an alternative to the analogue number hypothesis.

Certain potential economies and diseconomies associated with the possession of two forms of number representation may be identified. While in most circumstances digital number is likely to be relied upon in preference to analogue number, the existence of the latter may serve to provide an alternative strategy for computational contexts in which a fast and approximate result is needed and the precision which digital number affords can be sacrificed. In particular, analogue number may be utilized in circumstances when calculations on a natural analogue metric, such as size or intensity, or estimates of numerosity are called for. This would apply in particular when digital information is not immediately available.

The other side of the coin is that it may on occasion be a disadvantage to have an analogue number facility. We assume that digital and analogue representations are separate and effectively independent, since otherwise it would not make sense for the analogue system to be capable of coming to the rescue of an impoverished digital representation as just proposed. Circumstances can be envisaged when they would supply adequate but conflicting information about the numerical properties of a problem. Interference with performance would flow from the conflict as the invalid indications to the problem solution had to be suppressed. Normally this would be a matter of conflict arising from misleading analogue features of the problem, but the reverse situation could occur if inappropriate digital data were integral to the display of analogue problems. In practice, the relative speeds with which computation was effected on the two representations would determine whether or not interference between the two occurred. Of course, the possibility of a collaborative influence between the two representations should also be anticipated, emanating in performance as a facilitation effect. It will be recognized that this account owes much to the analogy with the Stroop and similar interference effects.

In fact, analogue number seems to have developmental precedence and there is substantial evidence that children use representations of number based on perceptual continua before they acquire digital number. Bryant (1974) refers to the well-established finding (e.g. Gelman 1972, Piaget 1968), that young children use length as a cue for number; they take the spatial or linear extent of a collection of objects as an external representation of numerosity. Gelman (1972) has shown that 4- and 5-year-old children use an absolute digital code for very small numbers of dots (two and three), while larger numbers tend to be handled using length cues. Moreover, Michie (1985) has noted interference effects arising from conflicting length cues in numerosity judgments. This work supports the hypothesis that there exists an analogue representation of number, from which digital number emerges initially for small numbers. However, it does not explicate the nature of an internal analogue representation, in which the magnitude of a number is modelled by analogy with a continuum not physically present, and for this we turn to studies of individuals in whom the representation of number has achieved stability and permanence.

The use of an analogue representation of number in adults was first suggested by Moyer and Landauer (1967). They discovered that the time required to choose the larger of a pair of digits decreases as a function of the logarithm of the difference between them. Further, for any given numerical difference, reaction times increase as the minimum digit increases. These two effects, the symbolic distance effect and the minimum effect, have been interpreted as indicating that numbers are internally represented as magnitudes, analogues to the representation of physical continua such as length and brightness. These representations of magnitudes are then compared by the same processes as are used in perceptual comparisons (Buckley and Gillman 1974; Foltz 1982; Moyer and Dumais 1978; Moyer and Landauer 1967, 1973). It is a feature of perceptual comparison processes that they act faster the greater the difference between the values to be compared, which gives rise to the symbolic distance effect. The minimum effect depends on this feature, and also on a non-linear encoding in which successive numbers appear closer and closer together in the magnitude representation, so that, for example, 7 and 9 are subjectively closer together, and therefore less discriminable, than 2 and 4.

There is some evidence that comparisons of multi-digit numbers proceed similarly. Hinrichs, Yurko, and Hu (1981) and Hinrichs, Berie, and Mosell (1982) found a symbolic distance effect in the comparison of 2- and 4-digit numbers, but the minimum effect was absent. Studies of the scale of psychological magnitude have been found by several investigators to be a power function (e.g. Schneider, Parker, Ostrovsky, Stein, and Kanow 1974) or a logarithmic function (Ekman 1964; Moyer and Landauer 1967; Rule 1969) of objective number. Galanter (1962) studied the psychological scale of monetary value and also found a power function, a result subsequently replicated by Kornbrot, Donnelly, and Galanter (1981).

While the availability of an analogue representation of number is well attested, its functionality is much less well established. The conditions under which it would be advantageous to use an analogue number system have largely not been examined experimentally. Kosslyn, Murphy, Bermesderfer, and Feinstein (1977), however, have produced evidence to show that analogue and digital representations can be evoked and processed simultaneously in a comparison task, and they reported that task conditions were critical in determining which route is the faster to provide a solution and which therefore emerges as the effective route. The lack of evidence on this point may be partly due to methodological difficulties in operationalizing the analogue/digital distinction. The first study reported here is an attempt to characterize further the analogue system, so that its behavioural manifestations can be better predicted and its functionality explored. It is concerned with the mapping function that translates objective number into its internal representation, and changes in this function associated with age. Children were studied in preference to adults, because the evidence seems to point to their greater use of analogue number. In Experiment 2 we examine the competing hypothesis that the mapping function observed in Experiment 1 is attributable not to the use of analogue number but to the reliance of the subjects on their biased experience of digital number.

Experiment 1

The first experiment examined how the number scale is calibrated, using a technique developed by Banks and Hill (1974) that produces a function relating external, objective number to the psychological scale of 'subjective' number.

The technique is very simple: subjects (Ss) are asked to generate a string of random numbers, as rapidly as possible, with no fixed upper limit, and in no special order. It is assumed that the form of the distribution of numbers obtained reflects the nature of each S's number scale. It is as though Ss refer to their 'internal rules' on which magnitudes are mapped, and select numbers from it in a random fashion. Intervals between numbers will then be represented in the output in proportion to their psychological size. Banks and Hill (1974) show how the assumption of random sampling allows a ratio scale to be derived from the subjective continuum. If each S's string of *n* numbers is rearranged from the smallest to the largest number, and at each ordinal position an average across Ss is taken, then these *n* averages estimate the expected values of the order statistic (Hogg and Craig 1970). Banks and Hill show that if sampling is uniform, the order statistics divide the line into *n* equally spaced points on the subjective scale, and give the *n* corresponding values on the objective scale. Thus a plot of the order statistic against the mean value of numbers given at each ordinal position represents the psychophysical mapping function.

Banks and Hill (1974) found with adult Ss in the range 1 to 1000 (numbers greater than 1000 were discarded, as being without meaning as magnitudes) that the subjective number scale is compressive—successive numbers lie closer and closer together. They found that a logarithmic function gave a good fit to the data over much of the range, but was less satisfactory for very small numbers. A possible resolution of this would be a combination of a linear scale for numbers 1 to 10 or 1 to 15 with a logarithmic one beyond. Banks and Hill also considered a power function with an exponent of about 0.67 as an alternative mapping function, and were inclined to prefer the power function, largely on the basis of comparability with psychophysical continua.

The authors were at pains to demonstrate that their technique did tap the subjective scale of number and not some irrelevant factor in the number generation task. One explanation which they were not able to rule out entirely was in terms of the 'spew' principle: that Ss produce numbers in proportion to the frequency with which they encounter them in daily life. Low numbers occur more frequently than high numbers in almost every application of number, from street addresses in *American Men of Science* to physical measurements (Benford 1938), and the decrease in frequency appears to be a logarithmic function of their size. Since the distribution of numbers generated was very close to a logarithmic function, Banks and Hill (1974) were not able to reject the frequency hypothesis as an explanation of the compressiveness of the number scale, though they showed that frequency effects do not give rise to non-linear sampling. They argue that frequency as a cause of non-linear number is not contradictory to their explanation. Benford regarded the log law as essentially the reflection of a more general law of nature. Nature, it appears, counts exponentially, and S's internal rules becomes calibrated as it is as a result of S's experience. In this sense, Ss' non-linear conception of number is seen as an analogue of nature. It is curious that, in spite of this, Banks and Hill stop short of endorsing a logarithmic analogue for the number scale. In fact, Moyer and Dumais (1978) cite this work as evidence for logarithmic number. We return to the question of frequency in Experiment 2.

The random-number generation technique was suitable for use with children and seemed particularly promising as a way of exploring changes with age in the representation of number. One modification to the procedure was introduced. Banks and Hill (1974) were concerned with the psychophysics of number, and their instructions implicitly directed Ss to operate with a scale. To avoid this orientation, and to render the task more intelligible to young children, the task was presented instead as an imaginary raffle draw. The child was asked to imagine himself or herself drawing the winning tickets from the hat. These instructions focus attention on the discreteness and nominal characteristics of number rather than their relatedness on a number scale. Under these conditions, a replication of Banks and Hill's results would be strong evidence for an analogue representation of magnitude.

The study has two sections. In the first, Ss generated numbers with no upper limit, and numbers over 1000 were eliminated prior to analysis. This allowed comparison with Banks and Hill's (1974) results, and across different age groups. In the second part, an upper limit of 100 was explicitly given, which allowed a more detailed comparison across age-groups.

Method

Subjects

Twenty Ss at each of four age-levels were drawn from schools in the Greater London area. Mean ages of the groups were 6.6, 8.7, 10.6, and 14.6 years. Each group was half male and half female.

Procedure

Under the 'no-limit' condition, Ss were instructed to imagine themselves running a raffle and drawing the winning tickets. They were reminded that there would be one ticket for each number, up to 'the highest number you can think of', and that the tickets were well mixed.

Ss' productions were recorded until they had generated 21 numbers up to 1000. Numbers greater than 1000 were disregarded.

Subjects who focused entirely on low numbers were invited to include higher numbers also. Those who gave strings of consecutive numbers were interrupted, and the task explained again. Some of the older Ss gave numbers in the way one would state a telephone number, e.g. '4 3 2 1', and were asked to use the form 'four thousand, three hundred, . . .'.

In the 'limit-100' condition, the upper limit was given in the instructions; otherwise, the procedure was exactly the same.

Results and discussion

Since the assumption of random sampling is crucial to the Banks and Hill (1974) analysis, Ss' verbatim productions were first scrutinized for evidence of non-randomness in the emitted sequence. Patterns of consecutive numbers, or of numbers given only in an ascending order, were taken to indicate that the numbers might not have been generated by means of a random-sampling rule. Across the four age-groups, there was reason to doubt that sampling was random in 22 cases in the no-limit condition and in 20 cases in the limit-100 condition. These Ss were excluded from further analysis.

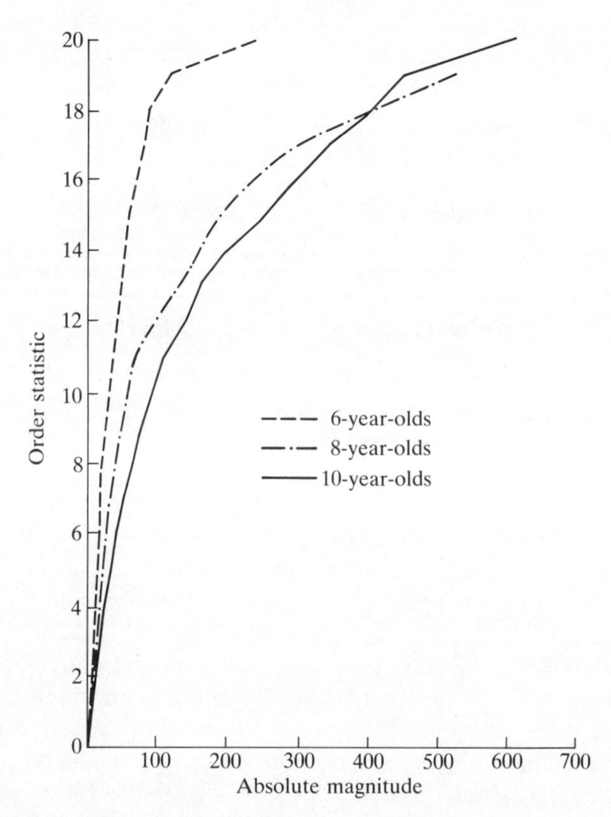

Fig. 8.1. Experiment 1: mean absolute magnitude produced in no-limit condition (14-year-olds omitted for clarity).

Following Banks and Hill (1974), the number sequences were ordered from the smallest to the largest, and geometric means at each value of the order statistic from 1 to 20 were computed for each age-group. Figure 8.1 shows the means (absolute magnitude) plotted against the order statistic in the range 1 to 1000 for the no-limit condition. The curvature indicates that the number scale is compressive in this range for all age-groups studied.

The best descriptions of these curves were determined by regression analysis, comparing linear, logarithmic, and power functions. The power

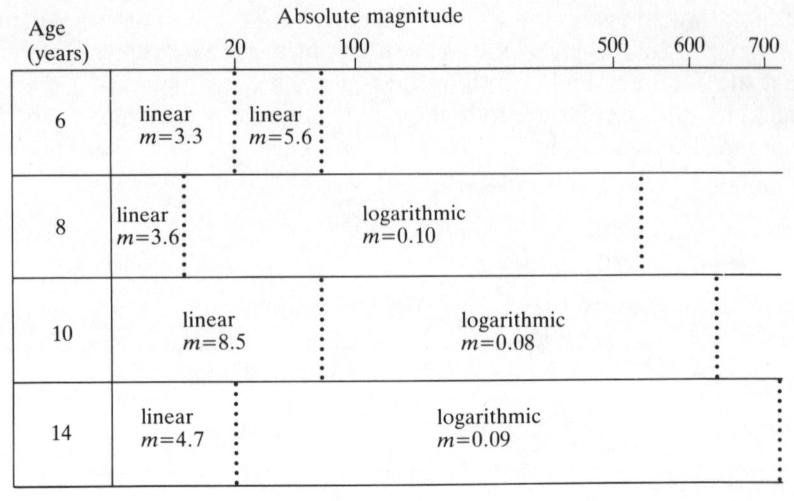

Fig. 8.2. Results of regression analysis on number productions in no-limit condition showing range and gradient(m) of mapping function.

function was found to be consistently a worse fit than the alternatives. Figure 8.2 shows the results of this analysis.

The 6-year-olds' data were best fitted by two intersecting linear regions: one for numbers up to 20, and a second, with different gradient, from 20 to about 60. The few data points above 60 were not adequately described by a straight line, nor by a logarithmic or power function. The change in gradient corresponds to a closer 'packing' or 'spacing' of numbers on the subjective scale above 20.

The remaining three subject groups' data are better understood as a combination of a linear representation of number at low numbers and a log function beyond. The numerical value which marks the transition point from linear to log varies between 8 and 14 years, but a computation on artificial data has shown that the location of this transition point is not accurately determined from data pooled from non-identical Ss.

Analysis of individual Ss' productions confirmed that a logarithm function was frequently the most appropriate for high numbers. Figure 8.3 shows the patterns which emerged. It is noticeable that where transitions occur between differently spaced linear regions or between a linear and a log region, the transitions appear at about 20 and at about 100. A discontinuity at 20 was also observed in the productions of some of the non-random Ss not included in this analysis. Particularly among 6-year-olds, there was a tendency to sample separately from above and below 20.

Figure 8.4 shows the mean absolute magnitudes produced in the 'limit-100' condition plotted against the order statistic. Again, Ss whose produc-

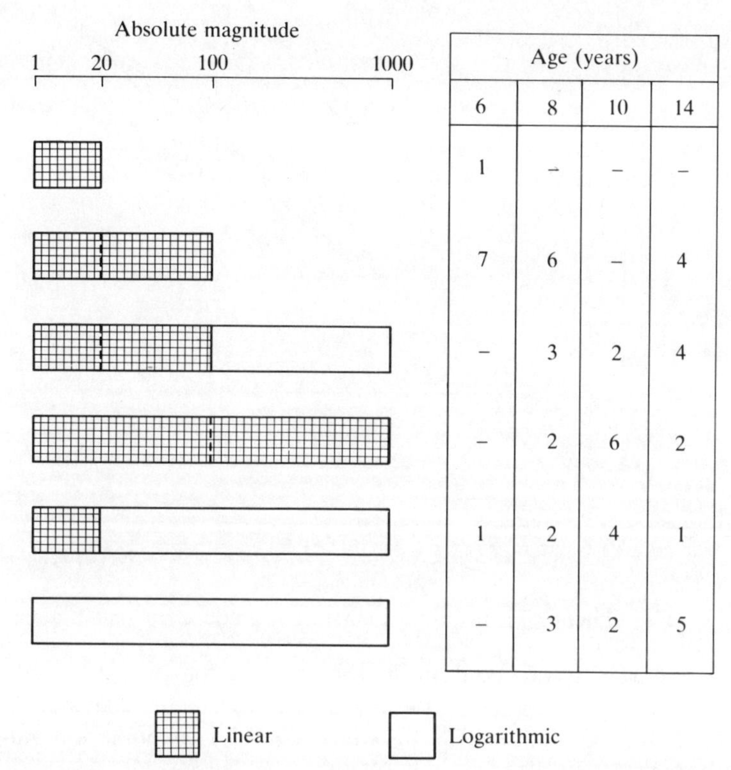

Age (years)			
6	8	10	14
1	–	–	–
7	6	–	4
–	3	2	4
–	2	6	2
1	2	4	1
–	3	2	5

Fig. 8.3. Types of number scale inferred in the no-limit condition with their frequencies of occurrence.

tions were detectably non-random sequences have been excluded. The curve for the 10-year-olds is now very close to a straight line, and the 8- and 14-year-olds' curves appear similar. As before, the most marked change occurs between 6 and 8 years.

These data were further analysed to determine the best description of the curves; the results are shown in Fig. 8.5. The 10-year-olds emerge as the only age-group to operate with a uniform linear scale throughout the range 1 to 100. The 8- and 14-year-olds show a discontinuity at about 20, where the upper linear region has approximately twice the gradient of the lower region. The 6-year-olds show a similar transition at 20 but their upper linear region does not extend as far as 100.

This study has shown two processes at work in the age range 6–14 years. The first extends the range of numbers which have an internal representation appropriate to the task. Large numbers (over 100) are apparently not available to the youngest Ss. When they emerge, their distribution reveals that they are represented as a compressive (logarithmic) function of their absolute magnitude; that is, an analogue representation. This result is

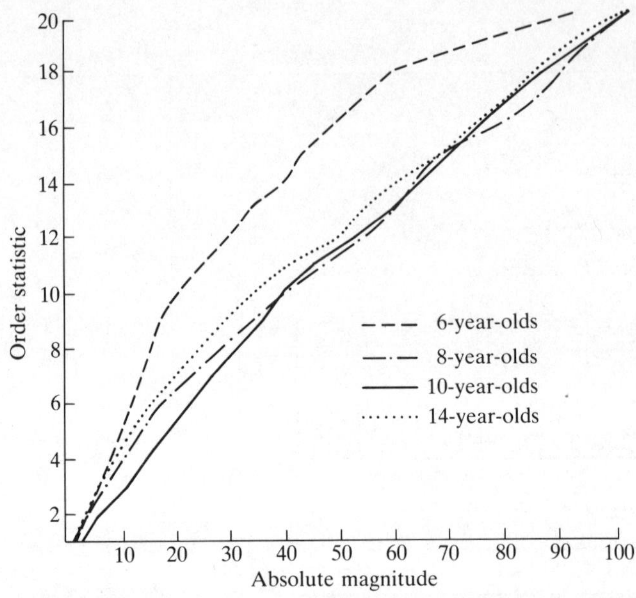

Fig. 8.4. Experiment 1: mean absolute magnitude produced in limit-100 condition.

in agreement with the contention that the analogue representation is ontogenetically prior to the digital. It seems likely that the distribution does in part reflect the frequency with which the child encounters number as magnitude.

Over the same age-range, a second process—digitalization—is at work,

<table>
<tr><th rowspan="2">Age (years)</th><th colspan="3">Absolute magnitude</th></tr>
<tr><th>20</th><th>60</th><th>100</th></tr>
<tr><td>6</td><td>linear
m=1.99</td><td>linear
m=5.00</td><td>logarithmic
m=0.096</td></tr>
<tr><td>8</td><td>linear
m=3.20</td><td colspan="2">linear
m=6.17</td></tr>
<tr><td>10</td><td colspan="3">linear
m=5.25</td></tr>
<tr><td>14</td><td>linear
m=2.85</td><td colspan="2">linear
m=6.29</td></tr>
</table>

Fig. 8.5. Results of regression analysis on number productions in limit-100 condition showing range and gradient(m) of mapping function.

and its effect is seen initially for small numbers. This process appears to have two components: unitization and abstraction. Numbers in the range 1 to 20 are unitized firstly, but their unidirectionality property shows that they are not fully abstracted from the concrete counting procedure for the youngest Ss. The range 20 to 100 emerges later as unit-based; a transition stage is evident where it is not commensurate with numbers 1 to 20. The smaller sized unit evident in the 20 to 100 range vestigially preserves its origins in a compressive representation, still evident among some subjects. Numbers above 100 appear to remain compressively represented for most subjects in this age-range.

There are indications that 10-year-olds have flexibility in the representation of number they employ, using an integrated digital representation of number up to 100 in the limit-100 condition and a linear/logarithmic combination in the no-limit condition. A similar effect of imposing an upper limit has been reported by Banks and Coleman (1981).

A final point concerns the 14-year-olds, who behaved in both conditions more like the 8-year-olds than the 10-year-old group. The implied regression may simply reflect the greater difficulty in securing a representative sample of subjects in secondary schools.

It could be argued, however, that the Banks and Hill (1974) paradigm invites the operation of an artefact; that is, subjects are in fact sampling from an array that is digital in form, but in which each entry is tagged by its experienced frequency, and the sampling from the array is weighted by the associated frequencies. It is assumed that the form of the relationship between experienced frequency and value is a negative exponential (as implied by Benford 1938).

It is immediately apparent that a single function of this form will produce a single functional relationship between the order statistic and value (unless it is assumed that there are fundamental differences in experienced frequency). In Experiment 1 (Figs. 8.2, 8.3, and 8.4) a variety of functional relationships was observed. Moreover, the form of the relationship should not alter because an upper limit is imposed on the sampling range, as again was observed in Experiment 1. Thus an experienced frequency explanation of the data is contraindicated by existing internal evidence.

Experiment 2

However, we should perhaps do better than this, and enquire about the nature of experienced frequency. Is it of an appropriate form to generate, say, a logarithmic function in a straightforward fashion? There appear to be no published sources for the incidence of numbers like those that are available for verbal material (e.g. Kucera and Francis 1967; Thorndike and Lorge 1944). Nevertheless, studies have appeared in the verbal learning literature which demonstrate that people are accurate and reliable judges of frequency of occurrence, and this applies to a dimension like word frequency (Underwood 1966) and even to aspects of verbal experience as obscure as phoneme frequencies (Carroll and Lamendella 1974). In the light of this, we obtained some data relating to Ss' judgments of the frequency of occurrence, in their own experience, of two-digit numbers.

Method

Each of 11 Ss provided ratings of a set of numbers chosen to span the range from 10 to 99. Lists were constructed so that each contained at least two doubles (i.e. of the form '*dd*' where '*d*' is chosen from the digits 1 to 9); two representatives of the form *d*0, *d*1 . . . *d*9; and at least two in the form 1*d*, 2*d*, . . . 9*d*. Four such lists enabled the range 10–99 to be covered. Extra representatives in some decades were necessary because single-digit numbers were not used and so the selection of material could not be symmetrical. For simplicity the ratings were on a three-point scale, with 'more often than the average' and 'less often than the average' as the end points, and accompanied by a line to imply equal distance between scale points. Most subjects completed one set of ratings in about two minutes. All subjects were asked to do all four sets, with an interval of a few minutes at least between each. Without a very large sample we cannot examine the relationship between judged frequency and numerical value in detail, but clear trends would be expected if we pooled in each decade, and biases with respect to particular values on the 'units' component will show if we pool across decades.

Results

The ratings for each subject within each decade were pooled, and again the data for each 'units' value were pooled across decades. Table 8.1 shows the average ratings for 11 subjects. Each mean is based on approximately 90 ratings.

There are a number of notable features of the results, not all of which bear directly on the analogue hypothesis, and we propose to focus on the most pertinent aspects of the data. It will be noted from Table 8.1, however, that the frequency profile within the units data shows two clear peaks at 0 and 5, and that, excluding these two values, all even values are rated higher than 2.0 and all odd values are rated lower than 1.9. In addition, a high mean rating is given for doubles. There was a highly significant degree of agreement among the Ss' ratings, with Kendall's coefficient of concordance for the units ratings taking a value of 0.48 (chi-square = 47.44 with 9 degrees of freedom; $p < 0.001$). As noted above, regularities in the units data would be imposed on the frequency function in such a way that they would tend to be obscured in the analyses conducted on the number generation data. But this does not apply to the data for the decade values.

Table 8.1. *Mean judged ratings as a function of numerical value.*

Value	Mean judged rating for 'tens' values*	Mean judged rating for 'units' values
0	—	2.52
1	2.82	1.79
2	2.39 (3.73)	2.11
3	2.22 (2.56)	1.83
4	2.12 (2.35)	2.10
5	1.94 (−1.39)	2.43
6	2.27 (1.60)	2.07
7	1.98 (2.67)	1.81
8	1.87 (1.39)	2.01
9	1.86	1.78

*Trend tests (z values) are reported in parentheses, centred on the middle of the three decades spanned by each test.

A plausible relationship between experienced frequency and decade value would seem to be a monotonically decreasing one. Moreover, if performance in the number generation paradigm is driven by experienced frequency, then, to judge from the form of the observed relationship between the order statistic and value, this is the underlying form of the frequency vs. value function. This general prediction is supported by an overall trend statistic calculated over the decades data ($z = 6.46; p < 0.001$). It is, however, apparent that judged frequency is not simply monotonically decreasing with value, since the downward trend is quite abruptly interrupted by a marked upturn in the 60s. To examine the trend in more detail, a series of trend tests was carried out for each set of three adjacent decades. Thus the trend for the 10/20/30s was calculated, followed by that for the 20/30/40s, and so forth. The results for this series of tests are reported in Table 8.1, where it can be seen that the downward trend has ceased for the 40/50/60s, is marginal at best for the 50/60/70s, is restored for the 60/70/80s, and is eroded again for the 70/80/90s. The secondary peak in the 60s seems quite clear since 6 of the 11 subjects gave their highest mean ratings for numbers in this decade.

Discussion

Thus although the general decreasing form of the frequency vs. value relationship is of the predicted form, there is an unexpectedly marked departure in at least one part of the range over which experienced frequency was examined. This effect seems strong enough to be reflected in the number generation data, but it was in fact not apparent in the analysis reported for Experiment 1; there was no indication of an irregularity in the function relating the order statistic to numerical value. We confess to a puzzlement relating to this outcome, since we have no explanation for it. The data were independently obtained by two of our number, and the form of the trend and the 60s reversal were apparent in both.

Despite the considerable plausibility of the frequency hypothesis, the data from Experiment 2 present a sufficiently strong indication that it cannot fully explain the form of the generated number functions from Experiment 1. In the event it turns out that the contribution of experienced frequency is differently and more complicatedly weighted than would be required to account for the number generation experiment. Nevertheless, we do not consider that this represents an exhaustive assessment of the hypothesis. On the assumption that experienced number is moderately stable, it would be appropriate to obtain data for number frequency judgments and for the number generation paradigm from the same subjects, in order to ascertain whether or not the detail of one is mirrored in the detail of the other. The present data indicate that the relation between the two would be imperfect, and this dissociation currently leaves the analogue number hypothesis intact. It should be borne in mind too that the analogue number hypothesis offers a parsimonious account of other findings reviewed above. In conjunction with conventional assumptions regarding cognitive processes (e.g. concerning automaticity, and the nature of perceptual interference in Stroop-type tasks), a theoretical attack on a quite broad range of findings and phenomena may be formulated.

General discussion

These two studies give further insight into the form of analogue number. The first study illustrates the duality of number and sheds some light on its resolution between 6 and 14 years. The higher numbers which have no meaning for most 6-year-olds emerge first in a compressive representation reflecting differential frequencies of exposure—an analogue of subjects' direct experience with everyday number. In this age-range, a second source is evident for the representation of low numbers. The counting procedure provides the origin of the unit-based digital representation of number. Progress appears to entail both integration, as initially distinct number-ranges become united, and differentiation: 10-year-olds can apparently adopt a different perspective on number depending on task demands. Digitalization itself appears to proceed through stages of progressive unitization and abstraction before number becomes properly symbolic.

This work gives a further clue to the source of children's difficulties with school mathematics. Formal arithmetic demands a fully abstract digital representation of number, yet this is seen to emerge only gradually from an analogue continuum, which itself derives from the everyday experience with number and value.

These studies give support to educationalists who argue that arithmetic skills should be taught with reference to relevant concrete situations. In addition, the random-number generation technique used in the first study may provide a simple diagnostic, to test for the maturity and extent of a child's digital representation of number, as an indication of this potential capability with respect to formal arithmetic.

References

Banks, W. P. and Coleman, M. J. (1981). Two subjective scales of number. *Perception & Psychophysics* **29,** 95–105.

Banks, W. P. and Hill, D. K. (1974). The apparent magnitude of number scaled by random production. *Journal of Experimental Psychology* **102,** 353–76.

Benford, F. (1938). The law of anomalous numbers. *Proceedings of the American Philosophical Society* **78,** 551–72.

Bryant, P. (1974). *Perception and understanding in young children: an experimental approach,* Methuen, London.

Buckley, P. B. and Gillman, C. B. (1974). Comparisons of digits and dot patterns. *Journal of Experimental Psychology* **103,** 1131–36.

Carroll, J. B. and Lamendella, J. P. (1974). Subjective estimates of consonant phoneme frequencies. *Language and Speech* **17,** 47–59.

Ekman, G. (1964). Is the power law a special case of Fechner's Law? *Perceptual and Motor Skills* **19,** 730.

Foltz, G. S. (1982). *The representation and processing of number.* Unpublished Doctoral Dissertation, University of Denver, Denver, Co.

Galanter, E. (1962). The direct measurement of utility and subjective probability. *American Journal of Psychology* **75,** 208–20.

Gelman, R. (1972). The nature and development of early number concepts. *Advances in child development and behaviour. Vol. 7* (ed. H. Reese). Academic Press, New York.

Hinrichs, J. V., Berie, J. L., and Mosell, M. K. (1982). Place information in multi-digit number comparison. *Memory and Cognition* **10,** 487–95.

Hinrichs, J. V., Yurko, D. S., and Hu, J. (1981). Two-digit number comparison. *Journal of Experimental Psychology: Human Perception and Performance* **7,** 890–901.

Hogg, R. V. and Craig, A. T. (1970). *Introduction to Mathematical Statistics.* Macmillan, New York.

Kornbrot, D. E., Donnelly, M., and Galanter, E. (1981). Estimates of utility function parameters from signal detection experiments. *Journal of Experimental Psychology: Human Perception and Performance* **7,** 441–58.

Kosslyn, S. M., Murphy, G. L., Bermesderfer, M. E., and Feinstein, K. (1977). Category and continuum in mental comparisons. *Journal of Experimental Psychology: General* **106,** 341–75.

Kučera, H. and Francis, W. N. (1967). *Computational analysis of present-day American English.* Brown University Press, Providence, RI.

Michie, S. (1985). Development of absolute and relative concepts of number in pre-school children. *Developmental Psychology* **21,** 247–52.

Moyer, R. S. and Dumais, S. T. (1978). Mental comparison. In: *The psychology of learning and motivation* (ed. G, H, Bower), pp. 171–255. Academic Press, New York.
Moyer, R. S. and Landauer, T. K. (1967). The time required for judgments of numerical inequality. *Nature* **215,** 1519–20.
Moyer, R. S. and Landauer, T. K. (1973). Determinants of reaction time for digit inequality judgments. *Bulletin of the Psychonomic Society* **1,** 167–8.
Piaget, J. (1968). Quantification, conservation and nativism. *Science* **162,** 976–9.
Rule, S. J. (1969). Equal discriminability scale of number. *Journal of Experimental Psychology* **79,** 35–8.
Schneider, B., Parker, S., Ostrovsky, D., Stein, D., and Kanow, G. (1974). A scale for the psychological magnitude of number. *Perception and Psychophysics* **16,** 43–6.
Thorndike, E. L. and Lorge, I. (1944). *The teacher's word book of 30 000 words.* Columbia University Teachers College Bureau of Publications, New York.
Underwood, B. J. (1966). *Experimental psychology,* Appleton, New York.
Wilden, A. (1972). *System and structure: essays in communication and exchange.* Tavistock Publications, London.
Wilder, R. L. (1968). *The evolution of number concepts: an elementary study*. Wiley, New York.

9

Levels of understanding and psychology students' acquisition of statistics

PHILIP T. SMITH

Introduction

The aims of researchers in the area of mathematical learning are often confused. Do we, as psychologists, want to study the retrieval of facts (such as $2 + 2 = 4$), the assembly of these facts in a complex situation (such as $583 + 432 = 1015$), or the more abstract understanding of the concept of addition? Do we, as educationalists, seek to develop the capacity for fluent and accurate calculation among our students (possibly at the expense of a thorough conceptual grasp of the problem) or would we rather have students who could sensibly discuss the issues in a given area of mathematics with only the haziest idea of how to carry out corresponding calculations? The reader may object that these are false choices; we want students who can calculate accurately *and* understand what they are doing, but it is sometimes difficult to achieve these aims simultaneously. Choice of experimental paradigm or teaching method inevitably focuses attention on some aspects of mathematical skill rather than others, and changes in theoretical emphasis, changes in the content of problems, and changes in technology all produce shifts in attention. Revolutions in physics often produce revolutions in the kinds of mathematics we want our students to know, emphasis on 'non-metric' methods of scaling or 'non-parametric' methods in statistics changes the types of calculations we want them to carry out, and the current range of choice of calculating-aid, from fingers to a main-frame computer, shapes the complexity of the calculations we expose them to.

The purpose of this paper is to discuss these overlapping and partially conflicting aims in mathematical teaching with reference to a particular context, that of university students' learning of elementary statistics. But I hope the issues discussed are sufficiently general for researchers whose interests are in other areas of mathematics learning to find parallels in their work.

Marr's levels

In Marr's book *Vision* (1982), he makes distinctions between levels of scientific enquiry, and these distinctions can usefully be applied in the present context. There are three levels; (1) computational theory; (2) representation and algorithm; (3) hardware implementation. For computational theory we wish to know: 'What is the goal of the computation, why is it appropriate, and what is the logic of the strategy by which it can be carried out?' With representation and algorithm we ask: 'How can this computational theory be implemented? In particular, what is the representation for the input and output, and what is the algorithm for the transformation?' And at the hardware implementation level we ask: 'How can the representation and algorithm be realized physically?' (All quotations from Marr 1982, p. 25.)

Marr uses these distinctions to clarify the structure of a scientific enquiry. In particular, he criticizes models of visual processes that are not fully motivated and do not rise above the level of algorithm. In his own theorizing, for example, he first proposes that surfaces are one of the primary constructs in visual perception. Then at the computational level he deduces from the known properties of physical surfaces what information, in the form of discontinuities in brightness, binocular disparity, etc., can usefully be employed to compute surfaces; at the representation and algorithm level he finds algorithms with attractive features (e.g. rapid convergence to a solution) which can compute surfaces; and at the hardware implementation level he discusses whether what is known about human physiology is consistent with this analysis.

In this paper I want to use Marr's framework, not as a model of scientific enquiry but as a model of a student's knowledge and skills. In order to do this I do not need to alter a single word in the quotations from Marr in the paragraphs above. The computational level, for the students' knowledge of statistics, is that of statistical principles; the representation and algorithm level is that of formulae and tables; and the hardware implementation is pen, paper, calculator, computer, etc.

Although it should be obvious from the preceding discussion, it is perhaps worth emphasizing that Marr's computational level refers largely to *concepts* not to *calculations*. For this reason, when I refer to arithmetic operations I shall call them calculations, not computations, even when they are carried out by a computer.

With Marr's levels of reference I believe we can put many issues in the acquisition of statistics into sharper contrast. For example, if we wish to discuss what concepts a student completing a course on statistics should have acquired, the debate should be conducted at the computational level. There are, for example, some techniques whose algorithms are of daunting complexity when it comes to carrying out calculations by hand, and which have not found their way into packages suitable for use on the microcomputers to which our students have access. However, sometimes these techniques are so important, either from a theoretical point of view or because of the impact they have had on psychology, that students really ought to know about them, even if they will never carry out a single step of the calculations. For this reason, at the University of Reading we teach randomization tests, multiple regression, and factor analysis on our second-year course to psychology students without the detail necessary for successful calculation.

At the algorithmic level we can discuss which algorithms the student should commit to memory (the degrees of freedom in a simple *t*-test?), which should be derived from general principles (the degrees of freedom in a multi-dimensional chi-square test?), and which should be retrieved from notes or textbooks (the degrees of freedom in a two-sample *t*-test, with grossly unequal variances?).

At the hardware implementation level we can discuss the relative merits of fingers, calculators, and computers, with the inevitable idiosyncratic adjustments to local conditions (the availability and reliability of particular microcomputers, the problems faced by a student with less than the full complement of fingers).

The individual points I am raising here are scarcely unfamiliar to mathematics teachers at any level. But what I wish to emphasize is that in principle these levels are independent: a commitment to teach the *t*-test, say, does not involve the necessary commitment to teach it at all of Marr's levels. Some students in some circumstances may actually benefit from being taught about the principles of the *t*-test without knowing how to calculate it.

There is a contrary view to this, which I suspect is quite widely held. According to this view we derive real understanding of a process only if we carry it out in detail ourselves. The prototypical case is motor skills: knowledge of a substantial body of a theoretical physics is no substitute for a little direct experience when it comes to riding a bicycle. This view has less merit with skills involving complex calculations: too often, in my experience, students cannot see the wood for the trees once the number of calculations becomes large. Thus, although at Reading we teach analysis of variance partly with the help of explicit calculations for simple one- and two-factor cases, this is intended more as an exercise which will illuminate the basic principles, not as preparation for the calculations involved in multi-factor analyses. These latter analyses we teach only via computer print-outs of the final ANOVA tables; i.e. shorn of any detailed calculation.

Studies of students' performance

So far this paper has been heavy on opinions and light on empirical evidence. Let me try to redress the balance by reporting a series of fairly

Table 9.1. *Correlated samples t-test problem: given these marks obtained by 8 students each attempting two essays, is there a significant difference in students' performance on the two essays?*

Student	Essay I	Essay II
A	6	8
B	7	9
C	5	7
D	8	7
E	9	9
F	8	9
G	5	7
H	6	7

informal studies of students' and colleagues' statistical abilities which illustrate my thesis.

Several of my examples are based on the following problem, shown in Table 9.1.

The only way students had been taught to handle this problem was via a correlated samples student's *t*-test. I shall call this the 'correct' method, despite doubts about whether such simple discrete data can be handled by a test that assumes sampling from a normal population.

In this case, t turns out to be 2.826, $\nu=7$. Tables give the critical value for t at the 5 per cent level (two-tail) as 2.365, so the result is significant at the 5 per cent level.

112 first-year psychology students at Stirling University attempted this problem, as part of an examination. They were allowed notes, books, and calculators. 100 of the 112 correctly perceived it as a correlated samples *t*-test, but their answers for t ranged from 0.086 to 25.3. What went wrong?

Interactions between hardware implementation and algorithm

63 per cent of the class admitted using calculators. But this group was not generally superior in its performance, despite having access to a more reliable device for carrying out arithmetic operations. While more students who used calculators obtained an answer correct to 4 significant figures in contrast to students who did not use calculators (11.3 per cent vs. 2.0 per cent), more students who used calculators failed to obtain the correct answer even to 2 significant figures in comparison with students not using calculators (46.7 per cent vs. 36.4 per cent). Another way of expressing this latter result is that if we arbitrarily define an implausible value for t in this question to be anything less than 0.75 or greater than 5.0, 29.0 per cent of the calculator users gave implausible values but only 16.2 per cent of calculator non-users. Alas, none of the differences quoted here is statistically signifi-

cant, but this does not detract from the main point: we do not convey immediate superiority to students in statistical tasks by improving their hardware. Failures at a computational or algorithmic level cannot be overcome by improvements at the level of hardware implementation.

As a more general point, I suspect that we will not create an intellectual revolution by giving everybody a home-computer, anymore than we can solve world poverty by giving everybody £100.

At the computational level, what many of my students lacked was a good intuition about what was a reasonable answer. We meet extreme examples of this lack of intution all too frequently: negative sums-of-squares in analysis of variance and correlation coefficients greater than unity are examples of impossible results that commonly appear in undergraduates' or even postgraduates' work. What I try to foster in my teaching is a more subtle intuition that works, for me, in the above t-test question something like this:

'There is a clear trend for essay II to have higher scores than essay I, but on such a small sample this may not be quite enough to obtain significance: if it is significant it will only be at the 5 per cent level. Since the critical value for significance at the 5 per cent level is 2.365, I expect t to be quite close to this, certainly within the range 1.5 to 3.5.'

I can then use this intuitive confidence interval as a check on my calculation, re-calculating t if it falls outside the confidence interval.

There is great variability in the degree to which psychologists develop these intuitions. I gave to several colleagues at Stirling University the t-test problem above, together with the information that the critical value of t for significance at the 5 per cent level was 2.365. I then asked: 'If you carried out a t-test on this data, what value would your answer have to be smaller/larger than before you would suspect you have made a mistake?' For three postgraduate students I obtained a median lower bound of 1.0 and a median upper bound of 5.0; for a mathematical psychologist the interval was narrower (2.0 to 2.7), commensurate with his thorough knowledge of the t-test. However, for a Research Fellow whose first degree was in linguistics, the interval was 1.5 to 20.0, and for a physiological psychologist who rarely analyzed data the interval was almost as large (3.0 to 15.0).

There are two lessons to be learned here: first, do not trust the statistics you read in linguistic journals (but I am sure you already follow this policy). The second point is more serious: the very people who are most likely to make mistakes in statistical calculations have the most lax criteria for accepting a solution as plausible. The proper reply to my question for someone who is inexperienced with the t-test is to say that any answer is likely to be wrong and should be re-checked. In fact they are saying almost the opposite: almost any answer is likely to be right.

Although I can offer no further systematic evidence on this point, my class-room experience suggests that this observation has some generality: weaker students have less grasp of what constitutes a plausible solution and are less inclined to check implausible answers. Greer and Semrau (1984) have also commented on how poor psychology students are at making approximate estimates, even in straightforward arithmetic tasks. The obvious teaching devices of encouraging approximate estimates to be made and demanding re-checking of implausible answers should be employed.

Interactions within the algorithmic level

As several contributors to this volume have pointed out, even the simplest of arithmetic processes can make substantial demands on the book-keeping and monitoring capacities of working memory. With the more complex calculations involved in a *t*-test, we might expect that even university students might have trouble assembling and interrelating the different parts of the algorithm. In fact it is possible to show that different parts of the same algorithm interfere with each other in competing for the limited resources of working memory. Or to use the more colourful analogy of Young, pandemonium created by different agents can lead to inaccurate performance.

Our Stirling students were taught to proceed through a precise set of steps, prescribed by Robson (1973, p. 78), in carrying out the *t*-test discussed earlier (e.g. step 1, calculate *d*, the difference between each pair of numbers; step 2, calculate Σd; and so on). This algorithm is shown in Table 9.2. One of the book-keeping tasks necessary to operate this algorithm is to keep track of where, for any given step, one has to go back in the calculation to retrieve the information necessary to continue the calculation. The number of steps intervening between the current step and the step from which information must be retrieved we call *retrieval depth*. Thus calculating Σd has a retrieval depth of 1 because one has to go back one step to retrieve information about the *d*'s necessary for the calculation. In Robson's algorithm, for example, calculation of d^2 has a retrieval depth of 3 (back to step 1 from step 4a); $(\Sigma d)^2$ a retrieval depth of 4 (back to step 2 from step 5a), and the final calculation of t a retrieval depth of 8 (back to step 3 from step 9).

It is not surprising that errors closely associated with keeping one's place are related to retrieval depth. For the 10 steps in the calculation where they could plausibly occur, I counted the number of errors due to inaccurate copying (e.g. substituting 484 instead of 448) or due to selection of the wrong component (e.g. substituting Σd instead of $\Sigma d/n$). The total number of such errors at each step was positively correlated with retrieval depth (Spearman's $\rho = 0.71$, $p < 0.05$).

Less expected was a positive correlation between retrieval depth and arithmetic errors. An arithmetic error we define as any error in basic

Table 9.2. *Robson's (1973) algorithm for carrying out a correlated samples t-test.*

Step	Formula	Calculation based on example in Table 9.1	Common errors
1	Calculate d	+2, +2, +2, −1, 0, +1, +2, +1	Ignore signs
2	Σd	9	
3	$\Sigma d/n$	9/8	
4a	Calculate d^2	4, 4, 4, 1, 0, 1, 4, 1	Trouble with negative squares
4b	Σd^2	19	
5a	$(\Sigma d)^2$	81	$(\Sigma d)^2 \neq \Sigma d^2$
5b	$(\Sigma d)^2/n$	81/8	
6	$\Sigma d^2 - (\Sigma d)^2/n$	71/8	
7	$\dfrac{\Sigma d^2 - (\Sigma d)^2/n}{n\ (n-1)}$	$\dfrac{71}{448} = 0.1585$	
8	$\sqrt{\left(\dfrac{\Sigma d^2 - (\Sigma d)^2/n}{n\ (n-1)}\right)}$	0.3981	Decimal point, $\sqrt{(10x)}$ instead of $\sqrt{x}$
9	$t = \dfrac{\Sigma d/n}{\sqrt{\left(\dfrac{\Sigma d^2 - (\Sigma d)^2/n}{n\ (n-1)}\right)}}$	2.8259	
10	$\nu = n - 1$	7	
11	Table-critical values	2.365	
12	Conclusion	Significant at 5% level	

calculation (addition, subtraction, multiplication, and division), including being unable to square negative numbers but excluding 'algebraic' errors (e.g. confusing Σd^2 and $(\Sigma d)^2$) and 'conceptual' errors (e.g. not realizing that the sign of d must be taken into account in calculating Σd). There is a positive correlation between the number of errors at each step and the retrieval depth of the step ($\rho = 0.83$, $N = 11$, $p < 0.01$). If we exclude one particularly difficult step (step 9, the only one involving calculating the ratio of two numbers each expressed as a decimal to several decimal places) then the correlation is marginally diminished ($\rho = 0.75$, $N = 10$, $p < 0.02$).

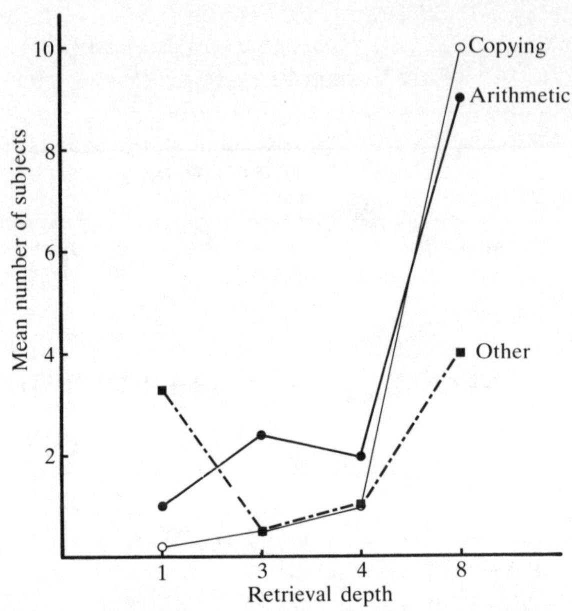

Fig. 9.1. Mean number of subjects (out of 100) making various types of error in the t-test algorithm as a function of retrieval depth. 'Copying' equals mis-copying information from an earlier part of the calculation or substituting the wrong information; 'arithmetic' includes all basic arithmetic errors; 'other' refers to algebraic and conceptual mistakes.

If we now look at the remaining errors that are not copying, mis-substituting, or arithmetical, there is no correlation between their number and retrieval depth ($\rho = -0.16$, $N = 11$, $p > 0.1$). These results are shown in Fig. 9.1.

The interest of these results is that they can be interpreted as suggesting that some but not all errors result from processes that are competing for the same limited resources in working memory. Having to keep track of which steps in a calculation are to be retrieved interferes both with 'book-keeping' errors such as mis-copying and mis-substituting and with basic arithmetic errors. The more difficult it is to keep track, as measured by retrieval depth, the more these other errors increase. In contrast, 'conceptual' errors arise from long-term memory storage or retrieval failure (a piece of knowledge is not available, for whatever reason, to the student): these errors are not dependent on the resources available in working memory and show no correlation with retrieval depth.

I am aware that this one small-scale study does not fully establish the importance of retrieval depth (more algorithms need to be studied, in particular algorithms carrying out the same calculations in different orders). The weaker point that this study has clearly established, however, is that arithmetic errors are not distributed randomly; so some features of the algorithm, perhaps described by retrieval depth, perhaps better described by some other concepts, are interfering with arithmetic accuracy.

The practical implications of this result are that if two parts of a complicated algorithm interfere with each other (keeping track interferes with arithmetic) we need to consider whether we can help students to obtain mastery of the task more efficiently by reducing or removing the arithmetic component in the early stages of acquisition. Once again this reduces to an argument for teaching computational theory and algorithmic skills separately from pure calculation.

Interactions between computational theory level and algorithm

In a more recent study at the University of Reading, I compared the performance of two groups of first-year students on an examination involving use of the Mann–Whitney *U*-test, the Wilcoxon *T*-test, and Spearman's ρ. The size of class necessitated teaching and examining them as two separate groups, and I exploited this situation as follows.

Both groups were given Mann–Whitney questions of comparable difficulty to provide a baseline from which to compare the rest of their performance. The useful feature of the other tests is that exactly the same data can be provided in questions where the examiner requires either the Wilcoxon test or a Spearman correlation to be carried out: one simply asks: 'Is there a difference between the two sets of data?' (Wilcoxon) or: 'Is there a correlation between the two sets of data?' (Spearman). Group C (the compatible group) received questions where (in the examiner's opinion) the phrasing of the question helped the student to select the correct test, Group I (the incompatible group) received questions where the phrasing was less helpful. In the example below the use of the labels 'before' and 'after' for the two sets of data was intended to favour selection of a Wilcoxon test (especially since in my classes I had given several examples of situations where experimenters might wish to know if there was a *difference* in performance before and after training). The phrasing thought to favour correlation selection involved sets of data labelled 'visual task' and 'auditory task', again because prior classes had used such labels in data illustrating correlations. An example of the type of problem is given in Table 9.3*.

The manipulations were successful. There was an interaction such that, relative to their performance on the Mann–Whitney tests, the incompatible group did worse on the Wilcoxon and Spearman calculations than did the compatible group (on arcsin-transformed percentage scores, $F(1,106)$ =

* Lest it be thought that manipulations of this sort in real examinations are unethical, I should point out that these manipulations made only a small difference to the overall marks, and all students in the class passed the class-work part of their course to which this examination was a contribution.

Table 9.3. *Example of problem involving the Wilcoxon T-test or Spearman's ρ-test: given these number of mistakes on a tracking task, before and after instruction, is there a significant difference between performance before instruction and performance after instruction (Group C, the compatible group), or is there a significant correlation between performance before instruction and performance after instruction (Group I, the incompatible group)?*

Subject	Before	After
1	1	2
2	2	2
3	1	4
4	6	5
5	4	9
6	11	13
7	6	7
8	7	10
9	7	9
10	7	6

4.98, $p < 0.05$). The decrement in performance was, however, small (mean scores were 83 per cent (Group C) vs. 81 per cent (Group I) for the Wilcoxon test; 87 per cent (Group C) vs. 82 per cent (Group I) for the Spearman test).

The point of theoretical interest is that the decrements, such as they are, do not simply arise from errors at Marr's computational level; i.e. choosing the wrong test: only 6 per cent of all subjects failed to select the right test. Rather, there are many errors associated with minor conceptual misunderstandings in the operation of the algorithms that appear slightly higher for the incompatible group. Such errors include ignoring sign when it should be taken into account, taking sign into account when it should be ignored, ranking zero when it should not be ranked, failing to rank zero when it should be ranked, etc. Many of these errors, as the preceding examples show, form complementary pairs: part of the algorithm appropriate only for the Wilcoxon test is being used for the Spearman test, and vice versa. Minor conceptual errors and omissions were made by 45 per cent of the subjects of the incompatible group on the Wilcoxon test (compared with 38 per cent in the compatible group), and the corresponding figures for the Spearman test are 46 per cent and 35 per cent.

Once again these differences are not significant, but once again a slightly weaker point is supported: a decrement in performance is not solely confined to the level at which it might be most expected (in this case, the computational level); rather, a general decrement is observed in the algorithmic level also. Less formally, if we confuse our students at one level, their performance may suffer at other levels as well.

Discussion

First let me admit that I am less than happy with the empirical basis of this paper. It ill-becomes someone who preaches the need for secure statistical analyses of properly controlled experiments to offer two rather informal studies, some of whose findings are not significant even at the 5 per cent level. While I could point to the converging evidence provided from different parts of these studies to bolster my position, I think it best to concede that the results can be regarded only as preliminary. Insofar as these studies are of value, their value derives from the questions they raise rather than the answers they provide.

What are these questions? Given that we know students' rate of uptake of new information is slow, that the number of activities they can monitor in working memory is limited, that most psychology students dislike statistics and statistics courses, that many of our colleagues, intelligent men and women though they be, regard statistics like a motorway driver regards fog (to be avoided if possible, otherwise to be driven through at reckless speed), what are we doing wrong and how can we put it right?

The message of this paper is to decouple. E. M. Forster's motto was 'only connect'; for elementary statistics teaching I am suggesting it should be 'only disconnect'. We fail to teach effectively when we throw a bundle of concepts, calculations, and computing altogether at our students, and I have tried to illustrate in this paper the sorts of negative interactions that can occur when we ask too much of the student. The tennis coach painstakingly builds a novice's service from component skills concerned with placing the feet, throwing up the ball, moving the racket, etc; so we statistics teachers might profit from a more component oriented approach.

In conjunction with several colleagues at Reading, in particular Elizabeth Gaffan and Alan Garnham, I have tried to develop a course where this decoupling is explicit. Our students are introduced in their first year to simple statistical tests, in a 'traditional' somewhat cook-book style, but in their second year they are given a term of lectures on statistical principles which contain hardly any calculations (Marr's computational level separated from the algorithmic level). This then leads on to a more advanced term of calculation, chiefly in analysis of variance. Use of computer packages, whether on department microcomputers or the university's main-frame, is also taught as a separate unit. A three-hour examination of the course is similarly structured, with questions testing verbal knowledge of experimental design and statistical principles separate from more traditional calculation questions.

It is difficult to evaluate this course objectively. I find it more interesting to teach than a traditional course. In particular the freedom to discuss any statistical principle, not just those that lend themselves to simple calculations, makes the course more intellectually stimulating. Certainly our better students are achieving a higher level of understanding of the subject. Certainly there are problems: the lack of consistency in styles and notation between lecturers, textbooks, and computer packages being the most obvious, though in principle this is easily remedied by rewriting the textbooks and the packages.

I am not offering panacea for all mathematics teaching. I am not suggesting even that the Reading statistics course offers the best way of teaching psychology students. What I am suggesting is, first, that there are several levels of understanding of mathematical skills and, before embarking on teaching such skills, we need to decide which levels of understanding we are most concerned to promote. Second, we psychologists in our laboratories know that human information-processing capacities are limited, and we should pay more attention to this insight not only in our laboratories but also in our class-rooms.

References

Greer, B. and Semrau, G. (1984). Investigating psychology students' conceptual problems in mathematics in relation to learning statistics. *Bulletin of the British Psychological Society* **37,** 123–5.

Marr, D. (1982). *Vision.* W. H. Freeman and Company, San Francisco, CA.

Robson, C. (1973). *Experiment, design and statistics.* 1st Edition, Penguin, Harmondsworth, Middlesex.

10

Understanding algebra

LAUREN B. RESNICK, EVELYNE CAUZINILLE-MARMECHE, and JACQUES MATHIEU

Introduction: the nature of mathematical knowledge

What does it mean to understand an algebra expression or an algebra rule? What role does understanding play in children's learning of algebra? These are the questions that motivate our research. We report on the early stages of a series of investigations that are simultaneously addressing some practical problems of learning mathematics and exploring fundamental issues in modern cognitive psychology: the nature of representational systems, the interaction between knowledge of principles and procedural performance skill, the ways in which knowledge changes as competence develops. While these issues can be studied in the context of almost any domain of human intellectual competence, they take on a special flavour in mathematics, for mathematical knowledge is in some fundamantal respects distinct from most other branches of knowledge. We begin, therefore, with a brief consideration of the special nature of mathematical knowledge and the psychological issues to which this characterization directs our attention.

Three features seem to us paramount in distinguishing mathematics from other domains of human knowledge. First, more than many domains of knowledge, certainly more than most domains that children must deal with, mathematics is concerned with abstract forms of knowledge. Second, mathematical knowledge is intimately linked to a specialized formal language that both imposes constraints on mathematical reasoning and confers extraordinary power. Third, the formal language of mathematics plays a dual role as *signifier* and *signified*, as the instrument of reasoning and as the object of reasoning. We briefly elaborate each of these features and their implications for learning mathematics.

Mathematics as abstract knowledge

In all domains of knowledge, forming a concept requires abstractions that go

The research reported here was funded by the Office of Naval Research (Grant Number N00014-79-C-0215) and the National Institute of Education. The opinions expressed in this chapter do not necessarily reflect the position or policy of the granting agency, and no official endorsement should be inferred.

beyond individual objects that can be denoted. The concept of a chair or a dog requires that children construct a representation that abstracts over the specific dogs or chairs they may encounter. However, in most domains there are at least *members* of a class or concept that one can point to, and it is possible to reason about specific cases. This is true even for so-called abstract concepts such as freedom, beauty, illness, and the like. We can define such concepts inductively, at least in a loose sense, by collecting examples of them. Mathematics does not have this property. There are not, strictly speaking, denotatable objects in mathematics. For example, although one can point to a set of three things, and to the written numeral 3, these physical objects do not in themselves have the property of number. Number is a strictly *cognitive* entity. People construct this cognitive entity, the concept of number, without benefit of any physical numbers to inspect or analyse. Yet number is the basic object of arithmetic. So we have in mathematics a domain in which, from the very beginning, people must reason about objects that exist only as mental abstractions.

Mathematics as formal knowledge

While one can reason about some aspects of quantity without using any written notation, there are very strict limits on how much reasoning about number one can do without using formalisms. From the moment that children try to quantify sets larger than 4 or 5 (the limit of the 'subitizing range'; cf. Chi and Klahr 1975) or put a measure on comparisons and estimations, formalisms that provide rigidly standardized labels and strictly constrain relationships among actions and objects are required. Take the apparently simple case of counting. Greeno, Riley, and Gelman (1984), building on Gelman's earlier work (Gelman and Gallistel 1978) on young children's understanding of the principles of counting, have analysed the knowledge that is involved in being able to use counting to quantify sets of objects. First, there is a standard set of labels (the count words) that must be used in a standard sequence with no omissions. While different language communities use different strings of count words (including some that use parts of the body, or names of familiar objects, rather than labels whose unique function is numeration), all counting requires absolutely strict application of conventional count words. There are no alternative labels, no connotative or contextual variations. Second, these labels must be paired with objects in accord with strict constraints (one label to one object, each used once and only once, in a fixed order). Throughout mathematics, then, performance and learning depend upon correct use of a formal system. This is true for very practical mathematical performances such as have been documented by students of 'mathematics in the streets' (e.g. Carraher, Carraher, and Schliemann 1985). However, the dependence of mathematical reasoning on formalisms becomes more marked as one proceeds to more complex levels of mathematical development. It becomes particularly evident when one considers algebra, where the formalism allows one to reason about operations on numbers without reference to any particular numbers.

The dual role of mathematical language

Throughout mathematics, the terms and expressions in the formal notation have both referential and formal functions. As referential symbols they refer to objects or cognitive entities external to the formalism. As formal symbols they are elements in a system that obeys rules of its own and they can function without continuous reference to the mathematical objects they name. Again first consider counting. The count words play a dual role. They are used both to name the individual objects in accord with the pairing constraints described above, and they are also used to refer to the cardinality of the *whole set* of counted objects. When one applies the count words in sequence to individual objects in the course of counting, one is using the count words as purely symbolic tokens in a formally constrained procedure. They do not refer to anything; they just keep the procedure running appropriately. However, when the same words are used to name the cardinality of the whole set that has been counted, they are names that have a referent, albeit a more abstract referent than many of the names in natural languages.

Meaning and symbol in algebra

The dual role of mathematical symbols is particularly obvious and complicated in the case of algebra. One great power of algebra is that it allows extensive manipulation of relationships among variables within a completely reliable system that does not require continuous attention to the referential meaning of the intermediate expressions that are generated. The fact that the algebra system can be 'run on its own', so to speak, is surely a factor in favour of its efficiency. Potentially unbearable demands on processing capacity would be placed on individuals who tried to reason through some of the complex problems for which algebra is used if, at each step, they were considering physical, situational, or specific quantitative referents for the transformations they produced. But algebra is not only a device for reducing capacity demands. Its very abstraction away from the situations, quantities, and relationships that are its referential meaning is part of what permits certain mathematical deductions to be made. To elaborate and explicate this claim it is first necessary to consider the meaning of an algebra expression.

What gives algebra expressions meaning?

Algebra can be seen as taking its meaning from three different sources. At one level algebra's meaning is completely contained in the formal system.

Expressions that are meaningful are those that are well formed within the formally defined system. Transformation rules are meaningful if they produce well-formed expressions when applied to initial expressions. At this level, expressions are meaningful if they can be derived from a set of axioms and postulates. The question of referential meaning for algebra expressions does not arise within the formal system itself. This purely formal level of meaning in algebra appears at the top of the triangle in Fig. 10.1.

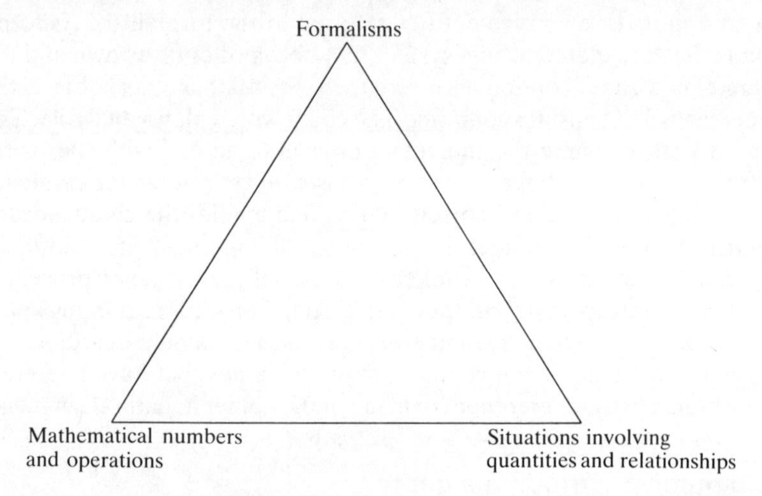

Fig. 10.1. Sources of meaning in algebra.

Algebra expressions and transformations also have referential meanings. The two kinds of reference that algebra expressions and rules can have are shown at the base angles of the Fig. 10.1 triangle. On the one hand, algebra takes its meaning from the world of numbers and operations about which it generalizes. Algebra expressions can be viewed as statements of the relationships that hold between numbers and operations on them *in general.* To show that an algebra equation is true or a transformation permissible, one can attempt to show that, in case after case where numbers are substituted for the unknowns in the expression, the equation still holds. Of course, such a demonstration cannot prove that in all possible cases the relationship will hold. However, for people who are fluent and comfortable with the world of specific numbers and operations on them, this kind of demonstration can do much to provide a psychologically convincing justification for algebra rules and expressions.

Algebra expressions and rules of transformation also refer to situations in which relationships among quantities and actions upon quantities play a role. Situations can be mathematized by expressing their quantitative relationships in an appropriate mathematical formalism. This is what we do when we solve problems by first writing appropriate equations. When this is done, it is sensible to think of the situation that gave rise to the equations as providing referential meaning for the equation. Once this referential function of situations is admitted, it also becomes possible to use situations to explain and justify the transformation rules of algebra. For example, the semantic equivalence of the following two stories justifies the sign-change rule in algebra; i.e. the rule that when parentheses are removed following a minus sign, the sign inside the parentheses must be changed $[a-(b+c)=a-b-c]$:

Mary had 30 candies. She gave 20 of them to her friends, 8 to Sandra, and 12 to Tom. (This story can be represented by the arithmetic expression $30-(8+12)$.)

Mary had 30 candies. She gave 8 of them to Sandra. Later she gave 12 candies to Tom. (This story can be represented by the expression $30-8-12$.)

These various meanings for algebra expressions and rules give rise to a paradox that is at the very heart of the learning challenge. On the one hand, formal expressions take their meaning in part from the situations to which they refer. On the other hand, algebra derives its mathematical power from divorcing itself from those situations. When algebra is used to solve problems, the problem-solver must first mentally represent the situation—the quantities, relationships, and transformations involved—and then write equations that encode those representations in algebraic notation. But the various transformations that are used in solving equations are possible in part because one stops paying attention to the specific situations that generated the equations and instead thinks only of the abstract operations and relationships that are also encoded in the formal expression. If the situations were not at least temporarily set aside, then some transformations would not make sense.

This point can be illustrated with respect to the most elementary algebraic expressions and the very first postulate that students learn in algebra. The law of *commutativity* says that $a+b=b+a$. This assertion is true when one takes as its referent mathematically defined numbers and operations (the left-hand base of the Fig. 10.1 triangle), but it is not always true if one takes situations as its referent. In terms of concrete reference situations there are actually two different kinds of addition. One corresponds to the *change* type of semantic situation identified by Riley, Greeno, and Heller (1983) for story problems (i.e. a starting quantity is modified by the addition or subtraction of another quantity). The other corresponds to the *combine* situation (i.e. two subcollections are combined to form a supercollection). These are shown diagrammatically in Fig. 10.2, using the conventions introduced by Vergnaud (1983). In the notation used, boxes enclose numbers that refer to states, and circles enclose numbers that refer to transfor-

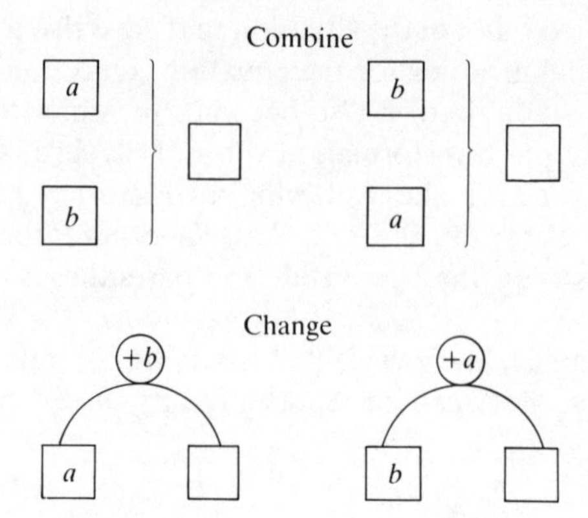

Fig. 10.2. the semantics of addition.

mations. In the combine situation, commutativity is apparent because the same kinds of mathematical entities, natural numbers, are combined. In the change situation, however, a natural number is modified by a directed number. In situational semantics, *a* changed by *b* is not the same as *b* changed by *a*. If we were thinking of the change situation, therefore, it would not be correct to say that addition was commutative. When we say addition is always commutative we *abstract* away from the specific situation, treat the quantities as pure numbers and addition as a mathematical operation.

The role of referential meaning in learning algebra

Substantial evidence exists that algebra learning is addressed by most school children as a problem of learning to manipulate symbols in accord with certain transformation rules, without reference to the justifications or meaning of these transformations. The best evidence is in the work of several investigators who have documented and analysed systematic errors in the kinds of equation manipulations that learners make. Carry, Lewis, and Bernard (1980), Greeno (1983), and Sleeman (1982, 1984) have studied the errors that students make in the early period of algebra learning (and that some people never overcome). Their work has shown that there is great systematicity in the appearance of errors; out of all the logically possible algebra errors only a small number are made with regularity. In this respect algebra appears similar to subtraction and other domains of arithmetic in which consistent errors have been documented (cf. Brown and Burton 1978; Hiebert and Wearne, 1985; Sackur-Grisvard and Leonard 1985).

The best-developed theories to date of how algebra malrules are invented are those of Matz (1982) and Sleeman (1982, 1984). According to Matz' theory, people generate malrules by constructing *prototype rules* from which they *extrapolate* new rules. Although the results are malrules, both the construction of the prototypes and the extrapolation come about through application of intelligent learning processes. An example appears in Fig. 10.3. The initial rule is the distribution law as it is typically taught in beginners' algebra courses. From this correct rule a prototype is created by generalizing over the operator signs. The prototype specifies not that multiplication can be distributed over addition, but that any operator can be distributed over any other operator. From this prototype, new, incorrect distribution rules can be constructed by substituting specific operations for the operator placeholders in the prototype. Sleeman attributes malrule formation to processes in which learned steps in a solution are forgotten or substituted for one another (manipulative errors) or to incorrect representation of algebra syntax (parse errors). Neither theory attributes to learners any representation of the *quantities and relationships* involved in algebra expressions and transformations. Instead, algebra malrules are interpreted strictly as deformations of symbol manipulation rules. A similar account of subtraction errors is given by VanLehn (1983).

1. The correct rule as taught:
$a \times (b + c) = (a \times b) + (a \times c)$
2. Prototype created by generalizing over operator signs:
$a \square (b \triangle c) = (a \square b) \triangle (a \square c)$
3. Incorrect rules created from the prototype:
$a + (b \times c) = (a + b) \times (a + c)$
$\sqrt{b + c} = \sqrt{b} + \sqrt{c}$

Fig. 10.3. The invention of an algebra malrule.

This tendency to construct inadequately constrained rules for symbol manipulation leads us to entertain the hypothesis that algebra learning might proceed more reliably if students were encouraged to relate the expressions and transformations of algebra to their conceptual and situational referents. These reference concepts and situations, to the extent that they themselves were well understood, could serve to constrain the kinds of modifications of taught rules that students would make. Algebra instruction rarely focuses on this problem of reference, except when teaching problem-solving as an application of algebra.

During the course of elementary schooling, one of the central learning tasks facing children in mathematics is constructing abstract concepts of number and operations. To do this, it seems likely that children link their knowledge about situations in which amounts are changed, combined, compared, etc., to their growing knowledge of how to quantify collections of objects and perhaps to their knowledge of certain spatial relationships (cf. Resnick, in press, b; Vergnaud 1982). Thus, knowledge about situations is already incorporated to some degree into the knowledge of numbers and operations that children must link to algebra formalisms as they move to the next stage of mathematics learning. Nevertheless, as we shall show, the task of linking situations to formal expressions, so that the expression can give meaning directly to the formalisms and the relationships they express, is one of considerable challenge for children in the early stages of algebra learning.

Principles for simple algebraic equivalences

The present paper reports on the first of a series of studies exploring the extent to which, despite the absence of much explicit help in developing a sense of the referential meaning of algebra, children may nevertheless develop that sense. In this study we limited our investigations to the most elementary algebra expressions and transformation rules. All expressions considered were three-term ones, involving only addition or subtraction. By choosing these very elementary aspects of algebra we were able to address the question of the kind of cognitive resources with which children initially approach algebra learning.

For the domain of expressions under consideration, understanding transformations and equivalences is supported by four basic principles; these are the principles that are often disregarded when children build malrules. They are:

Order irrelevance for addition This principle expresses the fact that it is permissible to add quantities in any order, regardless of how they are formed into subgroups (i.e. by parentheses) or their order of presentation. This principle is expressed in mathematics texts as two rules: commutativity and associativity. The subjects in our experiments more often expressed it in terms such as: 'You can do things in any order if you are only adding things in'.

Order relevance for substraction This principle is expressed in textbooks as noncommutativity of subtraction. Our subjects expressed it in terms such as: 'You can't change around subtraction' or: 'Starting with 8 and taking away 3 isn't the same as starting with 3 and taking away 8'.

Composition of quantity inside parentheses This principle expresses the fact that the two terms inside parentheses are the parts of a single whole quantity and that this whole quantity is to be added or subtracted according to the sign that precedes the parentheses. Alternatively, if the parentheses appear at the beginning of the expression, the composed quantity can serve as the 'starting set' from which the third term in the expression is to be added or subtracted. A composed quantity can also be compared or combined with another quantity. This principle can be used to explain the sign-change rules for removing parentheses.

Composition of transformations This principle expresses the idea that successive tranformations on a given quantity can be composed into a single, total transformation. For example, if b and c are both subtracted from a, as is indicated in the expression $a - b - c$, it is possible to compose the transformations into a single subtraction of the quantity $(b + c)$. This principle offers another means of justifying the sign-change rule. However, studies by Vergnaud (1982) and subsequently by Escarabajal, Kayser, Nguyen-xuan, Poitrenaud, and Richard (1984) show that the ability to compose transformations develops quite slowly, and that most children of 11- or 12-years-age cannot solve problems such as:

Peter won 6 marbles in the morning. He lost 9 marbles in the afternoon. How many did he lose altogether?

in which it is necessary to add two transformation quantities without knowing the starting quantity being operated upon.

Judging the equivalence of expressions

The study consisted of extended interviews with a small number of children at four different levels of the French *college*, serving children from about 11 years (*classe de sixième*) to 14 years (*classe de troisième*) of age.

The first part of our interviews consisted of a series of tasks designed to evoke children's knowledge about the equivalence of expressions. Three types of questions were used. Children were asked to: (a) indicate whether two expressions were equivalent or not and explain why; (b) state where parentheses could be placed in an expression without changing its value and explain why; and (c) state whether parentheses could be removed from an expression without changing its value and explain why. In some cases, subjects were first asked to judge expressions with letters, and then were shown numerical expressions of identical form and asked whether they would still make the same equivalence judgment.

The following expression pairs, shown first with letters and then with their numerical equivalents, provided opportunities for children to express knowledge of the order-irrelevance-for-addition principle:

$$a + (b + c)/(a + b) + c$$
$$a + (b - c)/(b - c) + a$$

The following expression pairs, shown with letters and then with numbers substituted, provided opportunities for children to express their knowledge of the order-relevance-for-subtraction principle:

$$a + (b - c)/a + (c - b)$$
$$a - (b + c)/(b + c) - a$$
$$(a - b) - c/(b - a) - c$$

Several types of items provided opportunities for children to express knowledge of the composition of quantity or composition of transformation principle. Two problems directly assessed children's knowledge of the effect of minus signs before parentheses:

Where can you put parentheses without modifying the result?

$$11 + 5 + 8$$
$$15 + 9 - 4$$
$$14 - 7 + 2$$
$$17 - 11 - 4$$

If you remove the parentheses, do you get an equivalent expression?

$a - (b + c)$ (and an equivalent numerical expression)

$a - (b - c)$ (and an equivalent numerical expression)

Other items required equivalence judgments for pairs of expressions in which parentheses were moved while maintaining the order of the letters and signs and their numerical equivalents. These sometimes violated the sign change constraint, as in:

$$a - (b + c)/(a - b) + c$$
$$a - (b - c)/(a - b) - c$$
$$(a - b) + c/a - (b + c)$$
$$(a - b) - c/a - (b - c)$$

Sometimes they did not violate the sign-change constraint, as in:

$a + (b - c)/(a + b) - c$ (and its numerical analog)

Other items were not directly tied to principles, but were included in order to examine the extent to which children understood constraints on exchanging positions of signs and numbers in expressions. These items also provided opportunities for children to express various malrules for symbolic expressions. Items of these types included expression pairs in which letter positions were maintained but sign positions were shifted:

$$(a - b) + c/(a + b) - c$$

There were also pairs in which sign positions were maintained but letter positions were shifted:

$$a - b + c/a - c + b$$
$$(a + b) - c/(a + c) - b$$
$$a - (b + c)/b - (a + c)$$

Details of responses to particular items and data on the relative difficulty of different types of judgments are reported in another paper (Cauzinille-Marmeche, Mathieu and Resnick 1985). Here we will concentrate on the broad features of children's reasoning about the expressions. We look especially for evidence of their attempts to use the principles to guide or justify their decisions about equivalence. We also try to characterize the ways in which the different kinds of meaning for algebra expressions—formal, conceptual, and situational—may interact.

Three strategies for judgment

The most striking finding from the interviews was evidence for three largely independent strategies that children used for making judgments about the equivalence of algebra expressions. These were (1) calculation; (2) rule-based evaluation; and (3) approximate evaluation. For the most part, these different strategies functioned as 'islands of knowledge' (cf. Lawler 1981), communicating very little with each other. This meant that children rarely used knowledge of one type to constrain or justify judgments of another type. On the other hand, one of the strategies would sometimes intrude on another to produce errors. We give examples in the following paragraphs.

Calculation

The preferred strategy of almost all the children, from 6e (sixième) (11 years) to 3e (troisième) (14 years), was calculation. Although we asked them to make judgments without calculating, the children calculated when they could. Children often replaced the letters in the literal expressions with numbers and then calculated, thus managing to use calculation even for literal problems. Sometimes the calculation proceeded *sub voce*, but the children would typically admit to having calculated when asked directly. Those who attempted a judgment without calculating often then calculated to check or confirm their judgments.

Children's use of the calculation strategy was reliable and largely correct. Thus, the calculation strategy represents children's reliance on the mathematics that they know best—arithmetic calculation. In principle, this reliance on calculation could serve the function of linking the algebraic formalisms to number and operation concepts (the left angle of Fig. 10.1). We would be inclined to believe that children were using calculation in this way if they had done things such as trying several different replacements for letters to see if a result remained constant. However, we did not observe this kind of behaviour; our subjects seemed to be calculating in order to deliver an answer rather than to explore relationships between quantities and expressions. On the other hand, our interview did not press children on this point, and since we appeared to subjects to devalue calculating, some of this kind of testing may have occurred without children's telling us.

Although the children's arithmetic was largely correct, there were some characteristic errors. These seemed to have two main sources: an incorrect understanding of the syntax of parentheses in the expressions, and intrusions into calculation of some incompletely learned formal rules for manipulation of expressions. The most frequent error observed was calculation from right to left instead of from left to right. This incorrect calculation rule led to a number of judgments that appeared to violate the principle of non-commutativity of subtraction. For example, a child in the 4e (13 years) judged $14 - (7 + 4)$ and $(7 + 4) - 14$ to be equal, saying: 'They will be the same. [In the first] it makes $7 + 4 = 11$, $11 - 14 = 3$, and [in the second] $7 + 4 = 11$, $11 - 14 = 3$'. This kind of 'backward' calculation was observed only when the expression had parentheses at the end. This regularity in conditions of occurrence, together with children's verbalizations, suggests that this error was the result of an incompletely understood rule that had been taught: 'Calculate the parentheses first'. Apparently some children took this instruction to mean that the entire direction of calculation should be reversed if the parentheses came at the right of the expression. As one child (4e) said for the problem $15-(6+2)$: 'I always start with the parentheses, and afterwards I continue with the number'. He then did $6+2=8$, followed by $8-15=7$. In both of the examples just given a characteristic linguistic inversion is present. The children said '11 − 14' instead of '14 − 11', and '8 − 15' instead of '15 − 8'. However, for these children and others who gave the 'parentheses-first' justification for their calculation, this appears to be a form of linguistic carelessness rather than a belief that the direction of subtraction is unimportant. This contrasts with a few of the youngest children who inverted subtraction in order to avoid negative numbers or 'impossible' subtractions. In such cases, the children might admit upon questioning that the inversion was not really legal. However, they were willing to relax the non-commutability-of-subtraction constraint in order to arrive at an answer.

Two other errors in calculation clearly derive from intrusions of formal rules of algebra manipulation. In one of these the child would first remove parentheses while (correctly) applying the sign-change rule, but then would calculate as if the parentheses were still present. For example, a child in 5e, given $14-(9+3)$, removed the parentheses to create $14-9-3$. However, he then behaved as if the parentheses around the 9 and 3 were still present. He subtracted 3 from 9, yielding 6, and continued by subtracting 6 from 14, giving the incorrect answer, 8. In a related type of error, children would sometimes mentally insert parentheses around the terms at the right of an expression. For example, for $14-7+2$, a child might calculate first $7+2=9$, then subtract the 9 from 14.

Rule-based evaluation

The strategy of rule-based evaluation consisted of applying formal rules in judging equivalence. Some of the rules thus applied were those that had been taught in school. Others were malrules invented by the children, apparently as deformations of the taught rules. In these invented malrules, we have evidence of the extent to which the formal system operates quite independently of what children know about mathematical principles and even about calculation. The errors made and the justifications given for them suggest that the children were for the most part attempting to learn algebra rules as purely symbolic manipulations.

Two of the most common errors concerned parentheses: in one case an over-estimation of their importance, in the second an under-estimation. In both, the function of the parentheses—to signal a composed quantity—was manifestly ignored by the children. The first error was to focus preemptively on parentheses, claiming that if the material inside parentheses was different, two expressions could not be the same. This led to judgments such as $a + (b - c) \neq (a + b) - c$. This malrule probably results from an intrusion into the formal system of the calculation rule that calls for operating inside parentheses first. Another such intrusion from calculation is evident when children judge as equal expressions that are left-to-right inversions of one another, such as $a - (b + c) = (b + c) - a$, justified in a typical case as 'the same because $b + c$ you have to do before a'.

The second common parentheses error seemed to derive from deformation of a formal rule rather than intrusion from calculation. This malrule claimed that the placement of parentheses was irrelevant, as long as the letters and signs retained their positions. Thus $a - (b - c)$ was judged equivalent to $(a - b) - c$; some children justified this equivalence by calling on the formal rule of associativity, misapplying it to subtraction. Another purely formal error was to judge it acceptable to commute signs as long as the letters remained in place. This produced judgments such as $(a - b) + c = (a + b) - c$. Finally, a special kind of formal error gave an improper segmentation of an expression without parentheses. It apparently results from searching for potential commutative pairs and treating them as if they had parentheses around them. For example, $a - b + c$ and $a - c + b$ were judged equivalent because $b + c$ commutes to $c + b$.

Approximate evaluation

This strategy, used only by some of the younger children and only for literal expressions, consisted of a kind of global reasoning about the effects of transformations (they either increase or decrease a quantity) without taking account of specific quantities. Here are some examples:

> $(a - b) - c$ and $a - (b - c)$. 'On the left you subtract, and then you subtract again. That will make a very little number. And on the right you do the subtraction $b - c$ and then subtract again, $- a$, that will make a very little number'.
>
> $(a - b) + c$ and $a - (b + c)$. 'On the left you subtract. You will get a number. Then you add c and you will get another, larger number. While on the right, it's going to be the opposite. You add b and c and get a number. And then you do $- a$. You will perhaps find the same number'.

It is not certain how these attempts at approximate evaluation should be understood. One interpretation is that children who use approximate evaluation are trying to bring to bear on the new domain of algebra their concepts of number and operations. However, new to algebraic notation, they treat the letters in the expressions not just as unknowns, but also as unknowable, and thus attempt to do calculations on indeterminate quantities. This interpretation would mean that the children were attempting to understand algebra in terms that went beyond pure symbol manipulation, that they were trying to make some referential sense of the new notation. But this may be too optimistic, for while approximate evaluation disappears among the older children, it is replaced not by a more refined linkage between the mathematical concepts and algebraic notation, but by purely formal evaluations. In other words, if approximate evaluation does indeed reflect children's initial attempts to understand algebra in terms of the mathematics they already know, this effort seems to be suppressed rather than used as a foundation for understanding in later instruction.

Knowledge of the principles

The preceding discussion has included comments on children's knowledge of the principles we had identified as justifying the transformational rules for our small subdomain of algebra. Nevertheless, it is useful to summarize here what our data on the judgment tasks suggests about the role of knowledge of principles in justifying equivalences and non-equivalences.

Order irrelevance for addition

Our subjects seemed to be well in command of this principle. They sometimes used the terms *commutativity* and *associativity* to justify equivalences in which the order of addition was inverted or in which parentheses were shifted in expressions involving only addition. More important, there were virtually no errors, for either numerical or literal expression pairs, in judging as equivalent expressions that involved only plus signs.

Order relevance for subtraction

This principle presents a much more complex picture. Many subjects, especially the older ones, explicitly called on this principle in refusing equivalences in which an inverted subtraction occurred. Nevertheless, as we have noted, subtraction inversion was a very frequent error. For some children, primarily the younger ones, this inversion seemed to be a direct violation of the order-relevance-for-subtraction principle. For numerical expressions, these children justified the inversion as something they had to do in order to give an answer when the written expression specified that a larger number be subtracted from a smaller. For literal expressions, they sometimes said that the order of subtraction does not matter. For other children, however, the apparent violation of the order-relevance principle appeared to derive from a misunderstanding of the syntax of the written expression. These were the children who believed that when there were parentheses at the right of an expression one interpreted the expression in a right-to-left direction. Given this belief, children who judged expressions such as $a - (b + c)$ and $(b + c) - a$ equivalent were not in fact violating the order relevance principle.

This distinction between true violation of the order relevance principle and apparent violation due to a misunderstanding of the syntax seems to us an important one. Children of 11- or 12-years-age who are willing to relax a fundamental constraint on subtraction (one they will admit to knowing, under questioning by the experimenter) in order to arrive at an answer are, we believe, expressing a fundamental belief about arithmetic. This is the belief that arithmetic is not a fundamentally lawful domain in which rules of procedure derive from the nature of numbers and their relationships, but rather an arbitrary domain in which rules are not related to fundamental principles. While our data is not extensive enough to test the hypothesis, it seems likely that such children would treat arithmetic learning in general as a matter of memorizing multiple rules for manipulating formal expressions (including numeric ones) rather than of reasoning about quantities and relationships among quantities. If this were so, we would expect them to develop and retain many malrules for both arithmetic and algebra, because they would be learning a complex, multi-rule domain, with few constraints on the kinds of deformations and modificiations of taught rules they might make. By contrast, children who invert subtraction because of a fundamentally syntactic error should be able to correct this error easily once the correct syntactic interpretation of parentheses at the end of an expression is pointed out to them. Furthermore, they may very well know and use a variety of basic principles in interpreting arithmetic and algebra expressions.

Composition of quantity and of transformations

These principles did not seem to play a role in the responses that our subjects gave to the equivalence judgment tasks. Parentheses were very salient for almost all the children. They knew that it was necessary to calculate inside parentheses first. Some children also knew the sign-change rule and applied it correctly to both numerical and literal expressions. We have also seen that children sometimes over-generalized the importance of parentheses and treated as non-equivalent any expressions that did not have the same material inside parentheses. Despite this salience, the only explanation for why one calculates the parentheses first or why it is necessary to change the sign when distributing a subtraction over the parentheses was in terms of rules for calculation—in other words, in formal arithmetic terms—rather than with reference either to mathematical concepts or to situations. That is, children might demonstrate that calculating without regard for parentheses yielded a different answer than calculating parentheses first, or that removing parentheses and changing the sign inside the parentheses and then calculating left-to-right yielded the same answer as calculating inside the parentheses first. None of the children ever justified these calculation effects in terms of composed quantities or composed transformations. Thus, there was no link made to mathematical concepts or to situations.

Linking algebra formalisms to the situational referents

It was striking that our children virtually never spontaneously used problem situations as a way of reasoning about the algebraic formalisms or the numeric expressions presented to them for judgment. Nevertheless, we wondered whether, when directly asked, they would be able to recognize links between situations and formal expressions and use those links to give referential meaning to algebra transformation rules. We therefore added to our interviews for most children a question in which the task was to make up a story that would generate a particular expression. In this part of the interview we used mainly numeric expressions of the same 3-term type as had been used in the judging exercises. In some cases, notably those in which children produced a first story correctly and without difficulty, we probed to see whether they could use the story situations to help them understand the reason for an algebra transformation rule. In each of these cases we focused on the sign-change rule. However, there were few such cases, for we found more difficulty than we had expected in generating stories to match expressions. The following sections describe and interpret the responses we encountered.

Successful reasoning about situations

To show how situations can be used to reason successfully about algebra transformation rules, we begin with the protocol of a child who clearly understood the rationale for the sign-change rule and was able to use story situations to reason about algebraic formalisms. Ger, a female student in 3e (about 14-years-old), was asked to make up a story for the expression 17 – 11 – 4. The text of her protocol appears in Fig. 10.4. To the right of the text there appears a schematic representation of the stories she composed

1 E: 17−11−4. Find a story for this expression.

2 S: At the start there are 17 poker-players. There's a settling of accounts. There are 11 of them who are very . . . and then. . . . Of those 17, there are 11 who are eliminated and there are 4 who come to put themselves on this list. . . .

3 E: What list?

4 S: On the list of the 11 who were killed off, so there are 4 who are also going to be eliminated, and then of the 17 there will remain . . . 2.

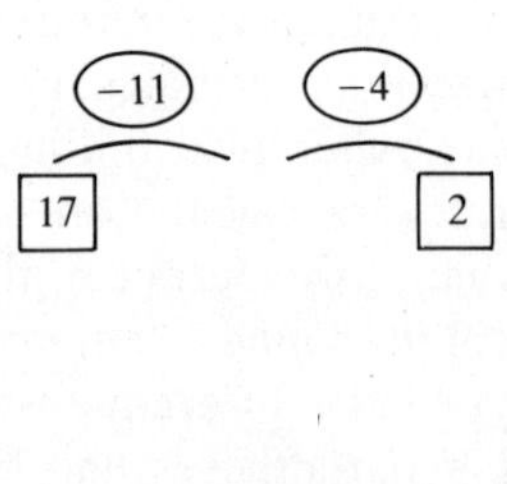

5 E: And what is the story one would tell for this one [17−(11+4)]?

6 S: We still have 17 poker-players, and to kill in cold blood 11+4 at one shot. . . . I don't see how you can do it. . . .

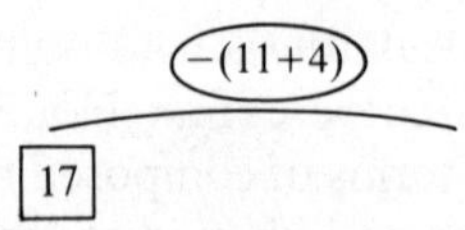

7 E: It's an interesting case, because you were asking why you change signs. In this poker story, don't you see the reason for the theorem? Here the story was 17 players; in the first round there are 11 players who leave and then 4. What can you say?

8 S: Here we have grouped, while there we have developed it step-by-step.

9 E: $a-(b+c)$. Can you remove the parentheses?

10 S: I will put $a-b-c$. You can take away the parentheses, but you get that.

11 E: Why? Can you tell me with your poker stories?

12 S: I have a certain number of poker-players. There . . . they are at the beginning. And I have some opponents. They are gathered to fight me, it's b and c. I arrive to fight them, b and c at the same time. So I can write that.

13 E: 15−(6+2).

14 S: The poker-players are 15 at the outset. They want to beat their most dangerous, most ferocious adversaries progressively. They start with the 6, the more difficult to beat than the others, and once they are eliminated they will attack the 2 remaining ones who are a little less dangerous.

15 E: $a-(b-c)$.

16 S: Always my poker-players. I have a certain number from which I remove. . . . I have my opponent, it's b, so I want to beat b. And c also wants to beat b. So it happens that in the fight he will be my ally to beat b. That may be why you have $a+c$, because with c, we two are trying to eliminate b.

17 E: 12−(7−1).

18 S: I have 12 poker-players, and 1 more, and they battle—the 13 against 7 poker-players, who are dangerous adversaries. That is they find themselves allied in combat to eliminate the 7.

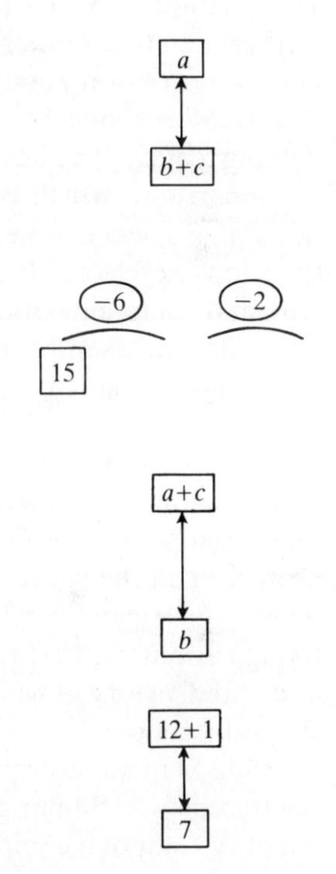

Fig. 10.4. Creating stories for formal expressions: Ger.

using Vergnaud's (1982) diagrammatic conventions for additive structures in arithmetic. Numbers enclosed in squares represent states, numbers enclosed in circles are transformations. This notation will prove useful in making clear the kinds of difficulties encountered by students less adept than Ger in constructing situational referents for arithmetic expressions.

Ger immediately (lines 2 and 4 of the protocol) constructed a story in which an initial quantity (the 17 poker-players) is transformed by two successive subtraction operators. This is shown in the first schema of the diagram. In contrast to cases we will describe later, Ger did not specify the intermediate state 6 that would exist after the first transformation. We take this as one indicator of her ability to compose transformations. Next she was asked what the story for 17 – (11 + 4) would be. Her response, in line 6, suggests that she recognized that the same basic situation is involved. This led the experimenter to point out that the stories could help her understand the sign-change rule. In line 8, Ger gave a fairly explicit explanation for this rule: she described the expression with parentheses as 'grouping' the transformations; the one without parentheses as developing them 'step-by-step'. The remainder of the protocol shows Ger using poker-players to reason further about parentheses and sign changes. Notice, however, that she did not always reason about transformations on a starting set. In lines 12, 16, and 18, she spoke of sets arrayed against each other (poker-players in fighting teams). The best mathematical representation of this seems to be as a comparison—which is one of the potential situational referents for subtraction. So, while poker-players were always involved, Ger's *mathematical* situations were actually of two types. The contrasts can be seen clearly in the schematic diagrams of the figure.

Our second example of successful reasoning is a child who seems to learn about sign change in the course of the interview. Man (male) was in the 5e (about 12-years-old). While children in this class had received some instruction regarding the sign-change rule, they had not mastered it. Man apparently was able to induce or reconstruct the rule by thinking about the relationships among quantities in a marble game. Figure 10.5 gives his protocol and the successive schematic diagrams. As is clear from the protocol, Man began (line 2) solving the parentheses problem that was posed by trying various calculations. Asked to make up a story, he immediately (line 4) created a story in which an initial quantity is transformed by two successive subtractions.

While Man did not specify the intermediate state, in lines 5 through 8 we see that he was thinking only of successive transformations on the original quantity—not of a composed transformation in which 15 is removed in two

1 E: Where can you put parentheses in 17–11–4?

2 S: (11–4) and (17–11). 17–11: 6, –4: 2. And 11–4: 7, 17–7: 10. . . . It's not the same! . . . You can only put the parentheses around (17–11).

3 E: Try to make up a problem about marbles with the expression 17–11–4.

4 S: I have 17 marbles. I lose 11 of them and I lose 4 more of them. I only have 2 left.

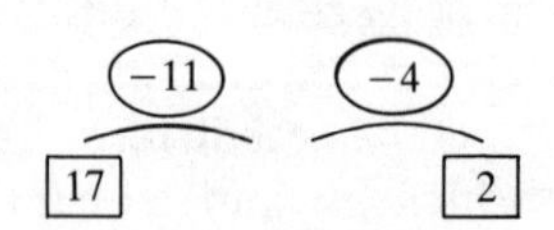

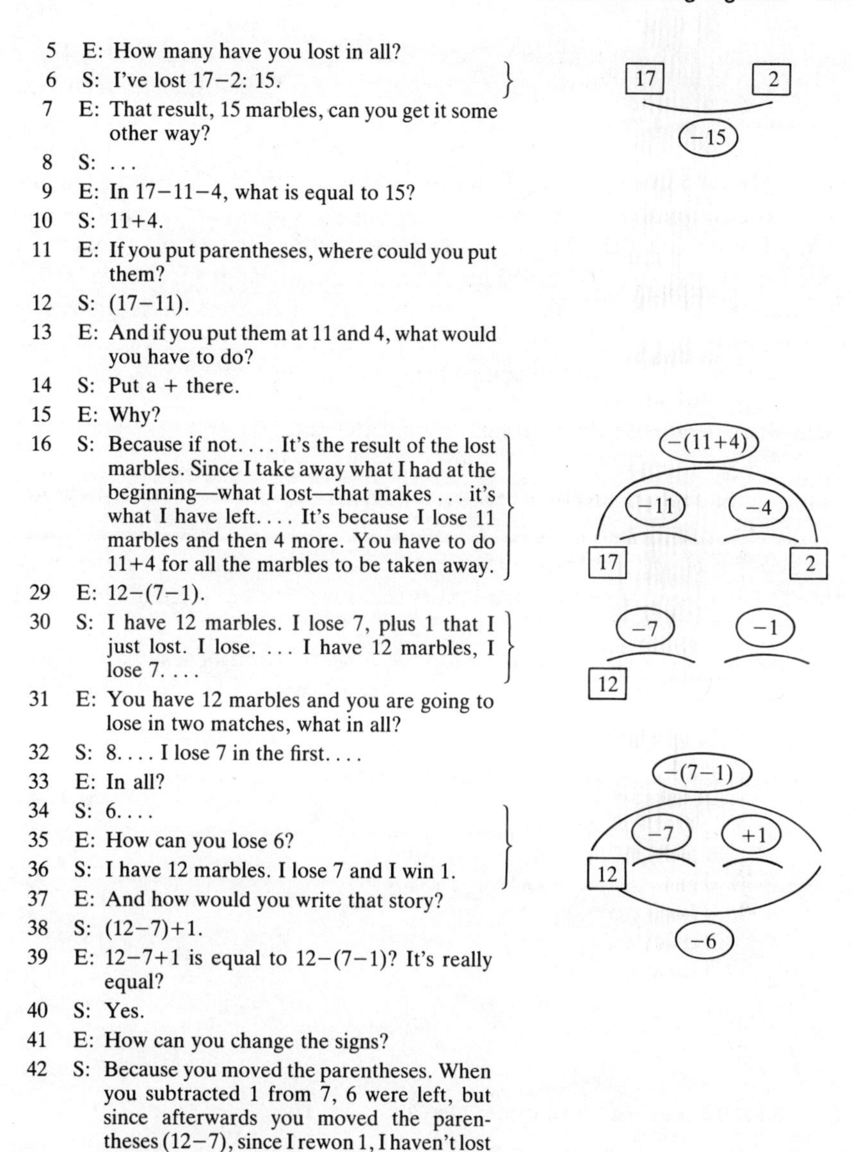

5 E: How many have you lost in all?
6 S: I've lost 17−2: 15.
7 E: That result, 15 marbles, can you get it some other way?
8 S: . . .
9 E: In 17−11−4, what is equal to 15?
10 S: 11+4.
11 E: If you put parentheses, where could you put them?
12 S: (17−11).
13 E: And if you put them at 11 and 4, what would you have to do?
14 S: Put a + there.
15 E: Why?
16 S: Because if not. . . . It's the result of the lost marbles. Since I take away what I had at the beginning—what I lost—that makes . . . it's what I have left. . . . It's because I lose 11 marbles and then 4 more. You have to do 11+4 for all the marbles to be taken away.
29 E: 12−(7−1).
30 S: I have 12 marbles. I lose 7, plus 1 that I just lost. I lose. . . . I have 12 marbles, I lose 7. . . .
31 E: You have 12 marbles and you are going to lose in two matches, what in all?
32 S: 8. . . . I lose 7 in the first. . . .
33 E: In all?
34 S: 6. . . .
35 E: How can you lose 6?
36 S: I have 12 marbles. I lose 7 and I win 1.
37 E: And how would you write that story?
38 S: (12−7)+1.
39 E: 12−7+1 is equal to 12−(7−1)? It's really equal?
40 S: Yes.
41 E: How can you change the signs?
42 S: Because you moved the parentheses. When you subtracted 1 from 7, 6 were left, but since afterwards you moved the parentheses (12−7), since I rewon 1, I haven't lost 7 but 6. I have to add it.

Fig. 10.5. Creating stories for formal expressions: Man.

steps of 11 and 4. The only way he could arrive at the total number lost was by subtracting his final quantity, 2, from the original 17. In lines 9 through 14, the experimenter led Man through a process in which he noticed that there is another way to find 15 and linked this to placement of parentheses. The experimenter's request for a justification of a sign change produced an explanation (line 16) that shows that Man was now thinking of 11 + 4 as a composed quantity, both parts of which must be subtracted from the original.

In lines 30 and 32, Man had trouble with a composed quantity that involved a negative number. He wanted to subtract both the 7 and 1 that are inside the parentheses. His response in line 32 shows further that he was mentally combining the 7 and the 1 to generate the number lost. Again, the experimenter's prompts helped him to construct the appropriate story (line 36), one that involved *winning* a marble even though the 1 has a minus sign in front of it—a considerable feat. His explanation in line 42 suggests that Man was grappling with the composition principles.

Difficulties in linking the situation to a formal rule

Things did not always go so well as they appeared to for Man. Several children showed themselves able to construct the appropriate stories but unable to use them to induce the rationale for the sign-change rule. The brief protocol for Hel (female, 3e) in Fig. 10.6 reveals a child more adept than Man at identifying a composed quantity that is subtracted from the initial set (line 4). However, Hel did not use that understanding of the situation to generate a proper formal expression. She used parentheses, but insisted (line 6) on keeping a minus sign in front of the 4, presumably associating the loss of 4 marbles with the need for a minus sign.

1 E: Make up a marbles story with the digits 17, 11 and 4.
2 S: Pierre has 17 marbles. He plays and loses 11 marbles. He plays again and loses 4 more.
3 E: How many marbles does he lose in all?
4 S: 15. 11+4.
5 E: How could you write that?
6 S: 17−(11−4).

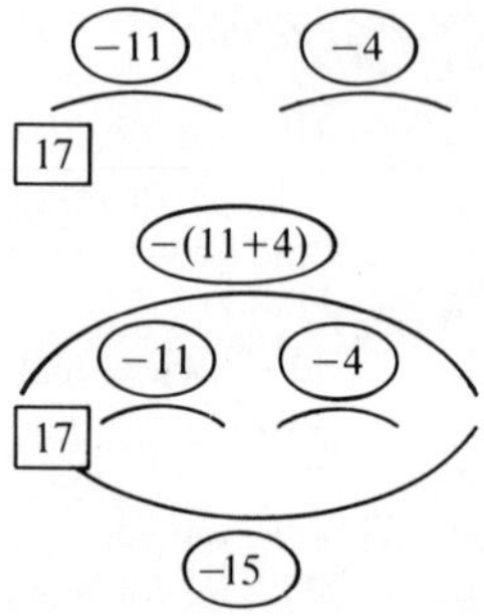

Fig. 10.6. Creating stories for formal expressions: Hel.

Hak (Fig. 10.7), a younger child (male, 6e), showed the characteristic preference for calculation as a way of solving the problem of determining placement of parentheses without changing the value of an expression (lines 4 to 6). Using this strategy, and perhaps recalling some class-room instruction on sign change, he successfully applied the sign-change rule and converted 17 − 11 − 4 into 17 − (11 + 4). However, he was not able to create a story to match the expressions. Instead, what he constructed at line 12 is a

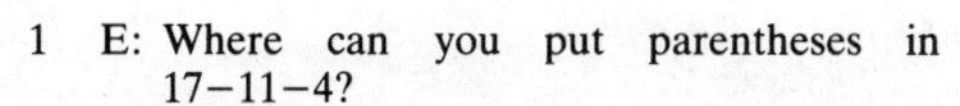

1 E: Where can you put parentheses in 17−11−4?
2 S: (17−11)−4.
3 E: Are you sure?
4 S: (17−11): 6. −4 equals 2. And 17−(11−4)= . . . 17−7=10. Oh no, that doesn't give the same result.
5 E: Could there be another way to do it, 17−(11 4), to have the same result, 2?
6 S: Oh, yes! +! 17−(11+4). 17−15=2.
7 E: So you do have the right?
8 S: Yes.
9 E: So 17−11−4 is the same thing as 17−(11+4)?
10 S: Yes.
11 E: Can you invent a problem where you would have to do those operations, 17−11−4? With marbles, for example.
12 S: 17−11−4. Well, I had 17 marbles, I lost 11 and I won 4 back? No?

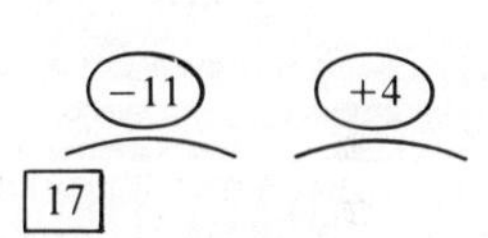

13 E: You have lost 11 and you win 4 back? How would you write that?
14 S: 17−11+4.
15 E: You had 17 marbles, you lost 11 and won 4. Do you write that like this, 17−11−4?
16 S: Oh, no. 17−(11+4), oh no, −4. I would lose 4 more.

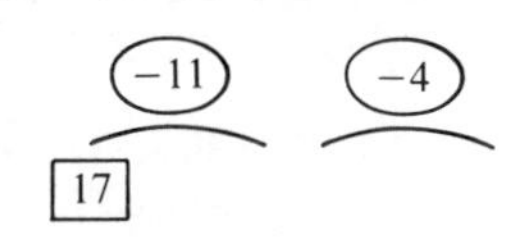

17 E: You would have 17 marbles, you would lose 11, and 4 afterwards. So how many marbles would you have lost in all?
18 S: 32.
19 E: You had 17 marbles, you lose 11, you lose 4. In all you lose 32?
20 S: Oh, how many do I have left then?
21 E: No, how many did you lose? You had 17. You told me that you lost 11, and then 4 more.
22 S: 15.

Fig. 10.7. Creating stories for formal expressions: Hak.

story with two successive transformations, one a subtraction, the second an addition. He thus merged the expressions with and without parentheses rather than confronting their semantic equivalence. At line 16, it appears that Hak caught on. He saw that in $17 - (11 + 4)$ he would *lose* 4. But his subsequent responses suggest that this reasoning was not very secure.

Difficulties in composing transformations

Difficulties in composing transformations seem to be at the heart of most of the difficulty that children had in linking stories to the formal arithmetic

5 E: Now you have 17−11−4, and you should try to find a problem where you will have to do 17−11−4 to find the result. For example, with marbles.

6 S: If Jean had 17 marbles and he . . . he lost 11 and afterwards he lost 4. And he only had 10 left.

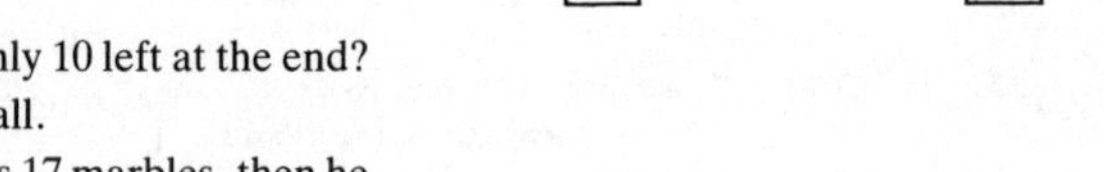

7 E: How is it that he has only 10 left at the end?

8 S: Because he lost them all.

9 E: At the beginning he has 17 marbles, then he loses 11, and then he loses 4, and you say that he has 10 left? (Draws diagram.)

12 S: Oh no! He would only have 2 left.

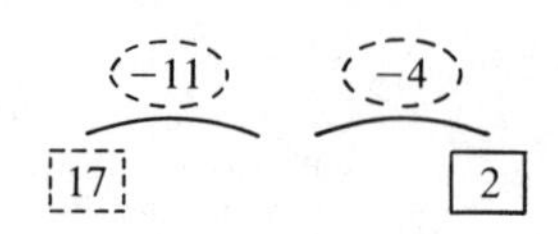

13 E: Yes. I agree that he would only have 2.

14 S: I made a mistake. 17−11−4, that makes 2.

15 E: How did you find it before? Did you find 10 by doing 11−4=7 and 17−7=10?

16 S: Ah . . . you have to start from *this* side, the left: 17−11=6, 6−4=2.

17 E: So, do you have the right to put in parentheses at the end: 17−(11−4)?

18 S: No.

19 E: Could you put in the parentheses at the end if you changed something? 17−(11 4)?

20 S: No, you can't do it because there is no sign.

21 E: Can you put in a sign to make it right? So you find 2?

22 S: . . . 17−(11+4). 15.

23 E: And what does the (11+4) represent with respect to marbles?

24 S: It's a number of marbles.

25 E: (11+4) is what marbles? You had 17 at the beginning, and (11+4) is what?

26 S: It's some other marbles.

27 E: What others?

28 S: I don't know.

29 E: We did the problem you said. You had 17 marbles, you lost 11, and afterwards 4.

30 S: He surely won back 4.

31 E: (Points to diagram above.) 11 and 4 is what you lost, and 2 is what remains. What is 11+4?

32 S: Well, maybe its marbles that he won . . . that he also lost. He only has two left.

Fig. 10.8. Creating stories for formal expressions: Cyrs.

expressions. The protocol of Cyrs (male, 4e) is instructive in this regard. Cyrs (Fig. 10.8) displayed a segmentation error of the kind discussed earlier in the context of the calculation strategy for evaluating equivalence. He told a perfectly correct story (line 6), but did his arithmetic without reference to the story. Instead he mentally inserted parentheses around 11 – 4, which led him to conclude there were 10 left at the end (see lines 15 and 16 for confirmation of this interpretation). This intrusion of a formal rule (calculate in parentheses first) into the story and arithmetic context is probably induced by the setting: Cyrs had just been asked to tell where parentheses could be placed in the intitial 17 – 11 – 4 expression. Whatever the reason,

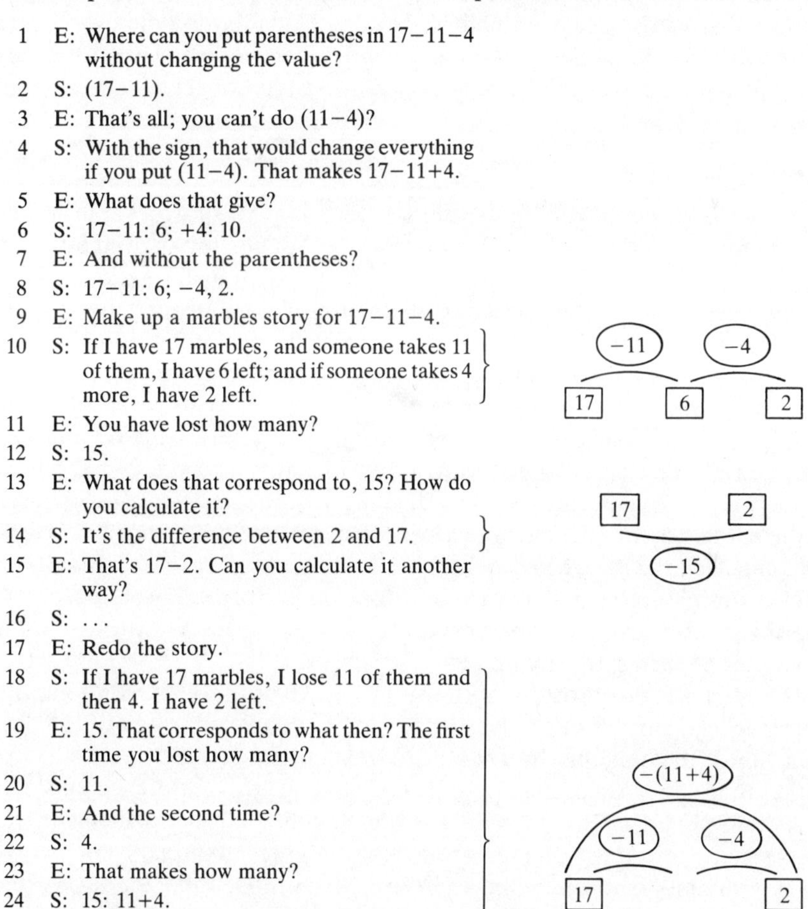

Fig. 10.9. Creating stories for formal expressions: Ali.

Cyrs' subsequent responses make it clear that he was not really able to link the arithmetic expressions to the story. In particular, despite considerable prompting and probing by the experimenter, he was unable to figure out what the subexpression $11 + 4$ might stand for in the story (lines 23 to 32).

Ali (female, 4e) was even less adept than Cyrs at composing the transformation. She made up a story for $17 - 11 - 4$ in which there were two successive transformations (Fig. 10.9). Evidence that she was not comfortable with the composition is that she gave an intermediate result, 6, after the first transformation (line 10). Further evidence is seen in the fact that she could find the total number lost only by taking the difference between 2 and 17 (lines 11 to 16). Ali eventually succeeded in composing the transformation, with considerable prompting by the experimenter (lines 17 to 26). However, we have no evidence as to how robust her final construction was.

Difficulty in composing any story

Each of the children considered thus far were able to construct appropriate stories for the initial arithmetic expression, although they showed different levels of competence in relating these stories to transformations of the formal expression. This difficulty seemed to centre around troubles in constructing or formally representing composed transformations. The children had no difficulty with non-parenthesized expressions in which successive transformations were adequate to represent the situation. However, to build a transformation into the story setting or to describe the relationship between the composed transformation in the story and the use of parentheses in the formal expression was difficult. A few of the youngest children in our sample demonstrated difficulties of a more profound kind. They could not, at least not without considerable help from the experimenter, construct even a straightforward successive transformation story that would generate the three-term arithmetic expression they were given. We present three examples of such difficulty here.

Dav's (male, 6e) protocol is shown in Fig. 10.10. In his first attempt to construct the story (line 2), Dav attempted a comparison situation ('6 less than him'). In doing this, he was able to treat both the 17 and the 11 as states rather than transformations. As will become clear, this preference for states was very strong for Dav, as for others who had difficulty constructing three-term stories. Another point to note is that Dav's first attempt did not involve all three numbers in the initial expression. This, too, will turn out to be a characteristic source of difficulty, indicating a fundamental problem with the notion of successive transformations. It also seems clear, even at this early point in the sequence, that Dav tended to treat the numerical expression as an instruction to do arithmetic. He subtracted and added, sometimes without reference to the sense of these operations.

1 E: What does that (17−11−4) mean? Can you make up a problem where you would do that? It doesn't matter what, with marbles. . . .

2 S: OK, there would be a boy who had 17 marbles, and then the other has 6 less than him. You would find his result and you would have 11. The other boy would have 11 marbles. To find that I would do a subtraction.

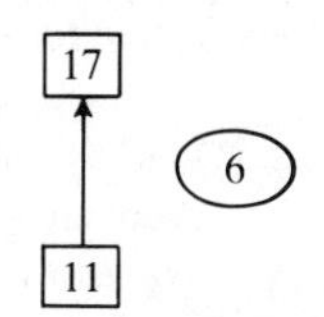

3 E: But to find 11 you did 17−6, and there we have 17−11−4, not 17−6. Let's start again. Boy 1 has 17 marbles. What can you invent to make 17−11−4?

4 S: The boy has 17 marbles. His friend would have 11 and another would have 4, just a little batch, and he would want to know . . . if you divided all that . . . if you subtracted all that, you would want to see how many are left if you do 17−11−4.

5 E: What does that story mean? You subtract one friend's marbles and another friend's marbles, that's strange. It would be better to take a marble game. You have some marbles and you play and you lose some marbles.

6 E: What would that give you? You have 17 marbles and you play. What happens in the first match, for example?

7 S: The first match, for example, he loses 6.

8 E: Why 6? That's not there.

9 S: At the second match, he has 11 left. He has lost 6.

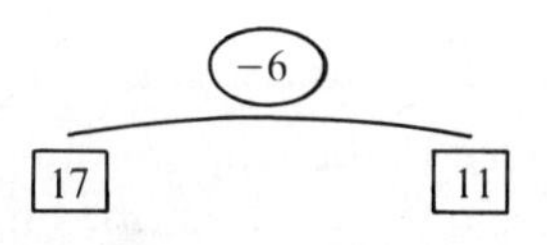

10 E: You always want to say that he has lost 6. But that 6, where are you going to put it? It's not there in 17−11−4.

11 S: Oh, yes, wait. In the first match he has 17. In the second he loses 11 and he would like to know how many he has lost.

12 E: So if he loses 11, you can make that 11. So you write that how? How many did he lose in the first match? How many does he have left now?

13 S: 17−11.

14 E: At the end of the first match, then, he has 17−11 left. And this 4?

15 S: At the end, when he goes home, he only has 4 left.

16 E: Oh, here we go again. And how are you going to write −4?

17 S: Well, you could add on the inside a little batch of 4. He has 4 left, and he would like to know how many more he lost in the second game.

Fig. 10.10. Creating stories for formal expressions: Dav. (*continued*).

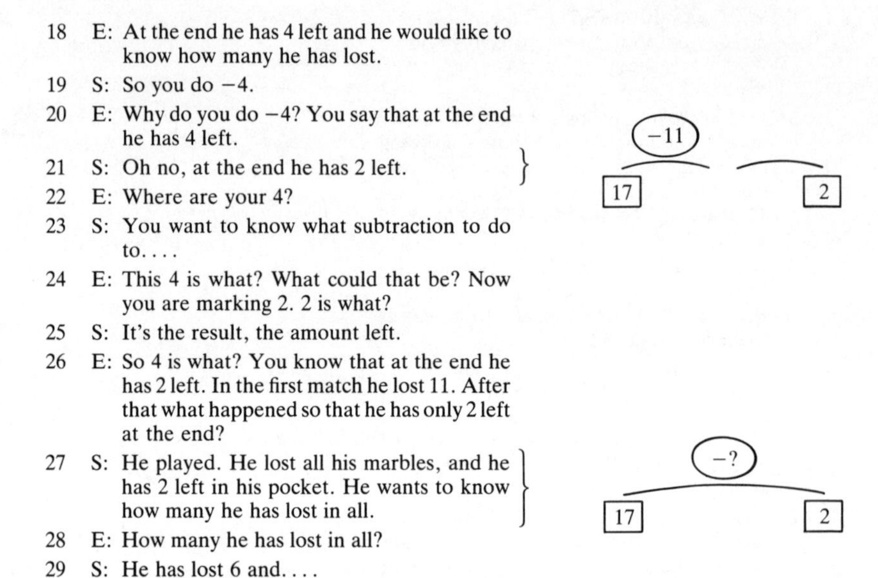

Fig. 10.10. Creating stories for formal expressions: Dav.

At the experimenter's request, Dav began again, and in line 4 produced a story in which all three terms were used, but each was used to describe a state, none was treated as an operator. Then he went on to subtract, as the expression prescribes, but without any meaningful reference to the story: since in the story nothing had been taken away, the question of how many are left is without meaning.

Next, the experimenter gave more prompts, and this finally led Dav to a transformation story (lines 7 to 9). However, he still wanted to treat the numbers given in the expression as states, and so used the difference between the first two numbers as the operator quantity. This allowed him to use the second number in the expression as a state. With some more prompting, Dav succeeded in treating the 11 as an operator (line 11). But he was unable to go on to treat the 4 as yet a second transformation, and instead used it as the number left at the end—a state (line 15). Despite prompts and questions by the experimenter, and an apparent momentary success in which Dav identified 2 as the number that was left (lines 21 to 25), he did not succeed in integrating the three terms in the given expression into a single coherent story. Throughout, he was driven by a search for subtractions to do and by the preference for treating the numbers in the expression as states.

Lau (male, 6e) shows a similar pattern (Fig. 10.11). He began with an incoherent story in which the numbers were not measures of the same quantity: the 17 referred to the number of boys, the 11 and the 4 to numbers of marbles (line 2). Prompted to let all the numbers refer to marbles, he used them all as states (lines 4 to 8), but even then had difficulty including all three in a single story line. Probably responding to the experimenter's request for a story with some action in it (line 9—'what happened?'), Lau began a story with a transformation (line 10), but he used the difference between 17 and 11 as the transformation number, keeping 11 as a state. At line 14, Lau succeeded in creating a transformation story, but he could not put the two transformations into relationship with one another. Instead, he created two stories, one for 17 – 11, the second for 11 – 4. Only with further prompting did he create a single, correct transformation story (line 20). However, note that he says it 'comes out to the same thing' as the previous story, probably meaning that the arithmetic is the same and not recognizing that the situation semantics are quite different. In the subsequent lines of Lau's protocol we see the same kind of difficulty with composing quantities that we have seen earlier for some of the more competent children.

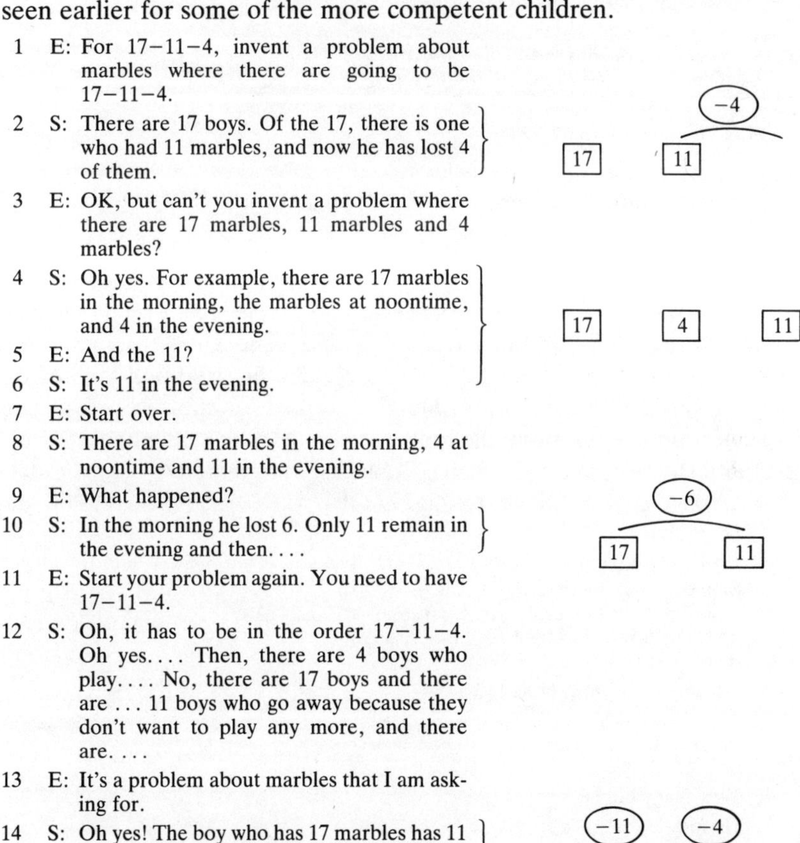

1 E: For 17–11–4, invent a problem about marbles where there are going to be 17–11–4.
2 S: There are 17 boys. Of the 17, there is one who had 11 marbles, and now he has lost 4 of them.
3 E: OK, but can't you invent a problem where there are 17 marbles, 11 marbles and 4 marbles?
4 S: Oh yes. For example, there are 17 marbles in the morning, the marbles at noontime, and 4 in the evening.
5 E: And the 11?
6 S: It's 11 in the evening.
7 E: Start over.
8 S: There are 17 marbles in the morning, 4 at noontime and 11 in the evening.
9 E: What happened?
10 S: In the morning he lost 6. Only 11 remain in the evening and then. . . .
11 E: Start your problem again. You need to have 17–11–4.
12 S: Oh, it has to be in the order 17–11–4. Oh yes. . . . Then, there are 4 boys who play. . . . No, there are 17 boys and there are . . . 11 boys who go away because they don't want to play any more, and there are. . . .
13 E: It's a problem about marbles that I am asking for.
14 S: Oh yes! The boy who has 17 marbles has 11 taken away; he has lost. And another boy who had 11 [marbles] lost 4 of them, –4.
15 E: It's two different boys?

Fig. 10.11. Creating stories for formal expressions: Lau. (*continued*).

16 S: Yes, it's two boys who play marbles. There is one who has 17 [marbles], he loses 11 of them and the other has 11 and loses −4.
17 E: How many marbles lost in all?
18 S: Counting both children?
19 E: Try it with one child who has 17 marbles.
20 S: Well that comes out to the same thing. There is a child, he has 17 marbles, he loses 11 of them, afterwards he loses 4 more.
21 E: How many does he lose in all?
22 S: . . .
23 E: You have 11 and afterwards 4. How many does he lose in all?
24 S: He loses 15.

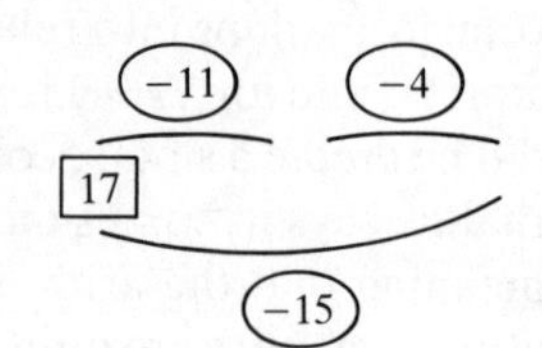

25 E: You have 17−11−4. You know that he loses 15. Can you write that another way than 17−11−4?
26 S: Yes. The boy who had 17 marbles lost 15 of them, but he lost 11 of them in the morning and 4 in the afternoon.
27 E: 15 is equal to what?
28 S: To what he lost.

Fig. 10.11. Creating stories for formal expressions: Lau.

Ste (male, 6e) is our final example (Fig. 10.12). He, too, preferred states to transformations. He initially treated both 17 and 11 as the number of marbles Paul has at different times (line 2), and he used only two of the numbers given in the expression. He also looked for subtractions to perform, even though his story did not demand it. Thus, he generated a 6 (presumably by subtracting 11 from 17), and at first treated it, too, as a state quantity. Starting over, Ste next used 17 and 4 (line 10), omitting the 11, and

1 E: Invent a marble problem with 17−11−4.
2 S: Suppose that there is Paul. He goes to school. He has 17 marbles. In the afternoon he comes home, he only has 11 marbles. So he only has 6.

17 11 6

3 E: He only has 11 left, or he lost 11?
4 S: No, he has only 11 marbles left. He had 17 marbles, he lost. . . . He comes home, he has only 11.
5 E: How many did he lose?
6 S: He lost . . . 6.
7 E: And then?
8 S: 6 . . . there is still the 4.
9 E: You have to invent a problem where there is 17, −11, and −4. Start your problem over.

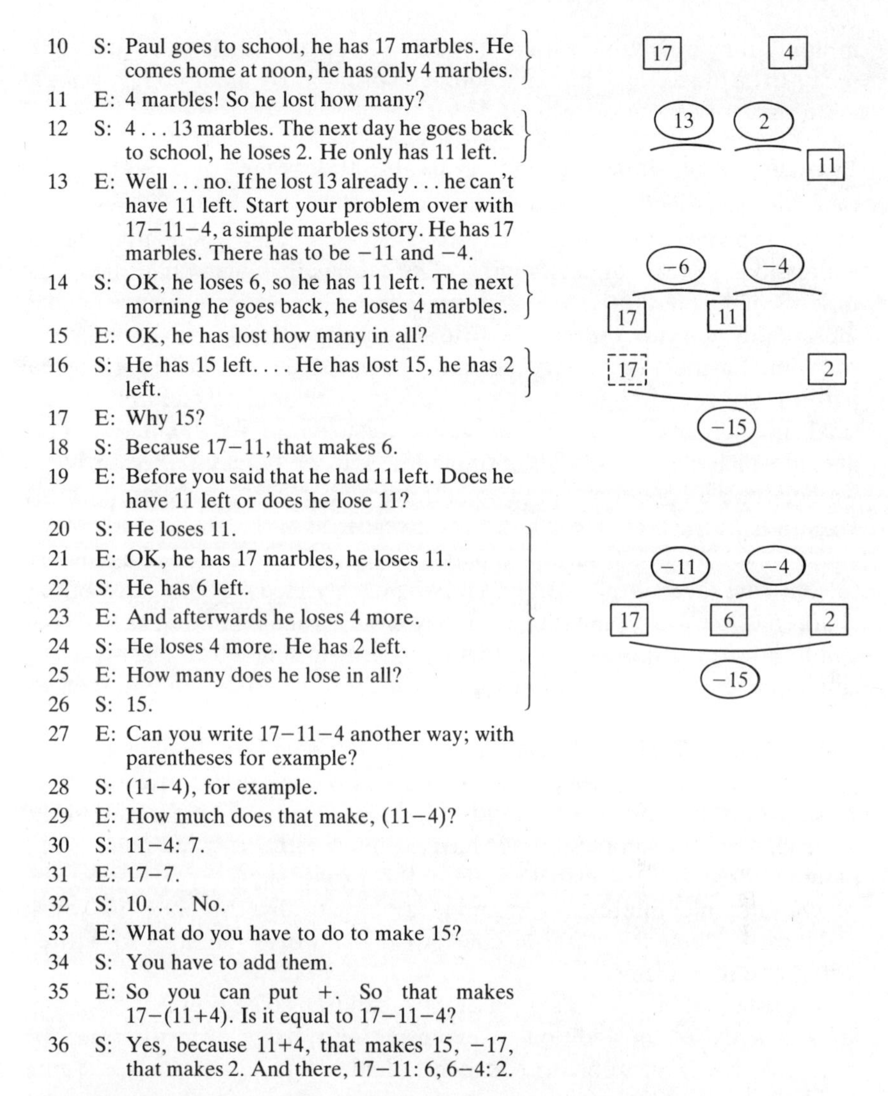

10 S: Paul goes to school, he has 17 marbles. He comes home at noon, he has only 4 marbles.
11 E: 4 marbles! So he lost how many?
12 S: 4 . . . 13 marbles. The next day he goes back to school, he loses 2. He only has 11 left.
13 E: Well . . . no. If he lost 13 already . . . he can't have 11 left. Start your problem over with 17−11−4, a simple marbles story. He has 17 marbles. There has to be −11 and −4.
14 S: OK, he loses 6, so he has 11 left. The next morning he goes back, he loses 4 marbles.
15 E: OK, he has lost how many in all?
16 S: He has 15 left. . . . He has lost 15, he has 2 left.
17 E: Why 15?
18 S: Because 17−11, that makes 6.
19 E: Before you said that he had 11 left. Does he have 11 left or does he lose 11?
20 S: He loses 11.
21 E: OK, he has 17 marbles, he loses 11.
22 S: He has 6 left.
23 E: And afterwards he loses 4 more.
24 S: He loses 4 more. He has 2 left.
25 E: How many does he lose in all?
26 S: 15.
27 E: Can you write 17−11−4 another way; with parentheses for example?
28 S: (11−4), for example.
29 E: How much does that make, (11−4)?
30 S: 11−4: 7.
31 E: 17−7.
32 S: 10. . . . No.
33 E: What do you have to do to make 15?
34 S: You have to add them.
35 E: So you can put +. So that makes 17−(11+4). Is it equal to 17−11−4?
36 S: Yes, because 11+4, that makes 15, −17, that makes 2. And there, 17−11: 6, 6−4: 2.

Fig. 10.12. Creating stories for formal expressions: Ste.

treated them both as states. The story became even more incoherent when he tried to use the 11 again as a state (line 12). Note, too, that Ste subtracted many times, regardless of the story line. Between lines 14 and 24, with much help from the experimenter, Ste finally generated a story with two successive transformations, 11 and 4. The experimenter then proposed that Ste now think about rewriting 17 − 11 − 4 with parentheses (line 27). Ste succeeded in doing this largely by using a calculation strategy to check a guess that seems to be induced by the experimenter's prompt. He made no reference at all to the story situation as a guide for the new construction.

Reasoning about situations: some general considerations

Our data on story generation cannot permit very strong conclusions about the range and extent of difficulties that children may have in relating formal expressions to their understanding of situations involving quantities. The task of composing stories was so novel that some children may have needed most of the interview time available to simply understand what was expected of them. Further, our interviews did not probe deeply enough to reveal patterns and relations among errors. Finally, our sample of children was too small and perhaps not representative enough to permit judgments about the ages at which particular kinds of difficulties may predominate. Neverthless, we believe we have uncovered something fundamental in children's early algebra learning that is worth deeper investigation.

We were, first of all, surprised that it was as difficult as it was for children to construct three-term stories. We had not expected this difficulty among children who were progressing normally in school mathematics, as all of our children were. If this kind of difficulty in linking simple formalisms to the situations represented by the stories is widespread among children, then we may have identified an important potential problem in learning mathematics that had not previously been suspected. The cases of Dav, Lau, and Ste are particularly surprising. We do not know how typical they are of 11- to 12-year-olds approaching the study of algebra. Two other children in the same class in our sample did not have as much difficulty, and none of the older children did. Nevertheless, these three cases suggest that it would be of considerable interest to examine systematically the ways in which children in elementary school come to relate story situations to written arithmetic formalisms.

A substantial body of research on how children solve simple story problems now exists. In addition to extensive data bases on error rates for different classes of problems (Carpenter and Moser 1983; Morales, Shute and Pellegrino 1985; Riley 1981), there are formal theories that account for these data (Briars and Larkin 1984; Riley *et al.* 1983), and analyses of the processes of linguistic interpretation in solving story problems (Bilsky, Blachman, Chi, Mui and Winter, 1986; Kintsch and Greeno 1985). These analyses all show that arithmetic story problem solution depends upon constructing mental models of the situations that constrain the relations among the given and unknown numbers. These models, rather than the particular words of the story, are used to infer the arithmetic operations that are required. For addition and subtraction stories, the models involved are the same ones we have considered here: transformations on starting sets (*change* problems); combinations of sets (*combine* problems); and comparisons of sets (*compare* problems). However, apart from Vergnaud's (1982) work, the story problem research has been limited to one-step problems, which would involve only two-term written expressions such as $(a + b)$ or $(a - b)$. Furthermore, story problem research has largely been directed at understanding how children solve story problems rather than in how they understand formal expressions.

Research on algebra problem-solving in adults and adolescents (e.g. Clement 1982; Hayes, Waterman and Robinson 1977; Hinsley, Hayes and Simon 1976) has concentrated more explicitly on the process of writing the equations specified by a problem statement. However, in most of this research it has been assumed implicitly that the difficulties lie in identifying relevant information in the texts and in determining the proper relationships among quantities and events, not in mapping these relationships onto formal expressions. Our results call this assumption into question, at least for beginners in algebra. The only research we are aware of that has directly examined children's ability to link mathematical formalisms (in this case, written arithmetic expressions) to familiar situations is the work of Conne (1984). This research, too, suggests that these links may be more difficult to establish than has generally been assumed.

Beyond the unexpected difficulty of the story construction task, our research thus far points to the probable utility of story construction in efforts to help children understand the referential meanings of algebra expressions and transformation rules. The case of Man provides direct evidence that it is possible to use stories to induce an explanation that makes sense of an initially arbitrary rule. We did not follow Man or any of our subjects to learn whether their brief exposure to thinking about formal expressions as having referential meaning had any effect on their subsequent learning. However, we were able to examine concurrent relationships between story construction ability and errors in judging equivalences among formal expressions. We classified children as mainly correct or mainly incorrect in their formal judgments and as able to construct appropriate stories without experimenter help, or not. Fifteen subjects were asked to make up stories for the expression $17 - 11 - 4$. Of these, three children were competent in both formal judgments and story construction and four were incompetent in both. Eight were able to make up the stories but incorrect in their formal judgments, but none were correct in their formal judgments but able to make up the stories. Thus, no one was successful in making formal judgments without also being somewhat competent at constructing stories. Although our data allow only rough judgments of competence, the pattern of relations between formal judgments and story construction competence allows us to put forward the hypothesis that interpreting expressions referentially is an important predecessor to successful learning of formal algebra.

In conclusion

We began this paper by pointing to a paradox. Situations, along with mathematical concepts of number and operations, are what provide referential meaning for algebra. Yet part of algebra's power derives from setting aside the referential meaning, especially the situational, and operating within the purely formal system. The research reported and discussed here has been devoted to discovering the extent to which children beginning to learn algebra are able to relate formal expressions to their situational and conceptual referents. But if algebra is so powerful as a purely formal system, perhaps this interest is misplaced, at least as a basis for instruction. Why, if algebra's power comes from its formal meaning, should any attention at all be directed to its referential meaning?

There are two reasons, we believe, for grounding algebra instruction in referential as well as formal meaning. The first, and most obvious, concerns the eventual utility of algebra as a practical tool for mathematical problem-solving. To say that situations provide referential meaning for algebra expressions is to say also that algebra expressions and equations can be used to represent the relationships that hold in situations. If one understands the referential linkage between situations and algebraic formalisms—and only if one understands this linkage—one is in a position to construct the equations that correctly 'mathematize' a situation and thus to use formal algebra to reach particular problem solutions. This being the case, it seems unlikely that instruction limited to the purely formal would be a good way to insure that algebra is learned in a way that will make it a useful tool rather than a purely formal exercise.

The second reason for grounding algebra instruction in referential meaning is less obvious. Teaching algebra formalisms as symbolic expressions of relationships that are otherwise well understood may provide a powerful set of constraints that will aid in the process of mastering the formalism. It is useful to think of malrules as resulting from a process of *constraint trading* that children engage in as they work to master the complex rule system that constitutes formal algebra. A good example of this is children's inversion of subtraction when judging the equivalence of algebra and arithmetic expressions. The children appear to know that order is relevant in subtraction. They also believe that answers must be generated whenever a numerical expression is given. In order to meet the second constraint, they 'trade away' the first, perhaps reasoning that in the special formal domain of algebra ordinary constraints from the world of numbers do not apply.

Our data and those of others who have studied the nature of errors in algebra learning suggest that this kind of incompletely constrained reasoning is quite common. Constraints implicit in the rules taught in school may never have been learned or may be violated in order to derive a rule that will allow one to give some response to the problem at hand. How can this tendency be combatted? One possibility is to attempt to make instruction much more explicit—so that all of the constraints are made explicit and children learn very precise conditions of application for each rule that they acquire. A number of proposals for improving algebra and other mathematics teaching *within* the formal system seek ways of doing this (e.g. Anderson, Boyle, Farrell and Reiser 1984; Lewis and Anderson 1985; Sweller and Cooper 1985). It seems likely that considerable improvement in formal algebra learning can be achieved in this way.

Yet the approach seems to overlook a fundamental characteristic of learning and instruction. In a situation where many rules, some of which differ in only small details, must be learned, it must be assumed that learners will play an active role in mentally constructing (or at least re-constructing) the rules which they come to apply (cf. Resnick, in press, a). It is in the nature of instruction, as it is of all communication, to be incomplete; the constraints that govern a rule or procedural domain can probably never be made entirely explicit. It is up to the learner to infer them—and thereby to construct a body of rules. Learners will engage in the necessary mental construction with whatever knowledge they have that they construe as relevant. This means that students' representations of the learning problem, along with the specific knowledge they have, will control the kinds of constructions they make. If learners believe that algebra consists of formal rules without referential meaning, they will use only what they know from the world of formal algebra, and perhaps the world of formal written arithmetic, to guide and constrain their rule constructions. If, however, they understand algebra expressions as having referential as well as formal meaning, they will be in a position to use what they already know about the semantics of situations and of fundamental mathematical concepts to constrain their formal constructions.

For these reasons it seems likely that if algebra is to be well learned by children, algebra expressions and laws of transformation must be related to the reference situations that might generate them, as well as to the mathematical constructs that they represent. At the same time, it will be necessary to keep in mind that to completely master algebra it is necessary to eventually ignore the reference situations and treat algebraic expressions as representations of relations that are independent of specific situations. The challenge of learning algebra, then, is both to relate the formalisms to the situations and mathematical principles that give them referential meaning, and to construct an understanding of algebra as a powerful formal system that contains its own internal meaning.

References

Anderson, J., Boyle, C. F., Farrell, R., and Reiser, B. (1984). Cognitive principles in the design of computer tutors. Paper given at *Sixth Annual Conference of the Cognitive Science Program.* Carnegie-Mellon University, Pittsburgh, PA.

Bilsky, L. H., Blachman, S., Chi, C., Mui, A. C., and Winter, P. (1986). Comprehension strategies in math problem and story contexts. *Cognition and Instruction* **3**(2), 109–26.

Briars, D. J. and Larkin, J. H. (1984). An integrated model of skill in solving elementary word problems. *Cognition and Instruction* **1**(3), 245–96.

Brown, J. S. and Burton, R. R. (1978). Diagnostic models for procedural bugs in basic mathematical skills. *Cognitive Science* **2**(2), 155–92.

Carpenter, T. P. and Moser, J. M. (1983). The acquisition of addition and subtraction concepts. In *Acquisition of mathematics concepts and processes* (eds. R. Lesh and M. Landau), pp. 7–44. Academic Press, New York.

Carraher, T. N., Carraher, D. W., and Schliemann, A. D. (1985). Mathematics in the streets and in schools. *British Journal of Development Psychology* **3,** 21–9.

Carry, L. R., Lewis, C., Bernard, J. E. (1980). *Psychology of equation solving: an information-processing study.* University of Texas Press, Austin, TX.

Cauzinille-Marmeche, E., Mathieu, J., and Resnick, L. B. (1985). *La coordination des connaissances: micro-mondes et genèse des règles de résponse—etude réalisée à propos de l'appropriation par de jeunes éléves des premiers concepts de l'algebre elementaire.* Unpublished manuscript. Laboratoire de Psychologie Genetique, Université Rene Descartes de Paris V, France.

Chi, M. T. H. and Klahr, D. (1975). Span and rate of apprehension in children and adults. *Journal of Experimental Child Psychology* **19,** 434–39.

Clement, J. (1982). Algebra word problem solutions: thought processes underlying a common misconception. *Journal for Research in Mathematics Education* **13**(1), 16–30.

Conne, F. (1984). Une épreuve de calcul. *Interactions Didactiques, 6.* FAPSE Genève et Seminaire de psychologie, Université de Neuchatel.

Escarabajal, M. C., Kayser, D., Nguyen-xuan, A., Poitrenaud, S., and Richard, J. F. (1984). Compréhension et résolution de problèmes arithmétiques additifs. *Les Modes De Raisonnement.* Association Pour La Recherche Cognitive, Orsay, France.

Gelman, R. and Gallistel, C. R. (1978). *The child's understanding of number.* Harvard University Press, Cambridge, MA.

Greeno, J. G. (1983, August). *Investigations of a cognitive skill.* Paper presented at the Annual Meeting of the American Psychological Association, Anaheim, CA.

Greeno, J. G., Riley, M. S., and Gelman, R. (1984). Young children's counting and understanding of principles. *Cognitive Psychology* **16,** 94–143.

Hayes, J. R., Waterman, D. A., and Robinson, C. S. (1977). Identifying relevant aspects of a problem text. *Cognitive Science* **1,** 297–313.

Hiebert, J. and Wearne, D. (1985). A model of students' decimal computation procedure. *Cognition and Instruction* **2**(3 & 4), 175–205.

Hinsley, D., Hayes, J. R., and Simon, H. A. (1976). From words to equations: meaning and representation in algebra word problems. In *Cognitive processes in comprehension* (eds. M. Just and P. Carpenter). Erlbaum, Hillsdale, NJ.

Kintsch, W. and Greeno, J. G. (1985). Understanding and solving word arithmetic problems. *Psychological Review* **92,** 109–29.

Lawler, R. W. (1981). The progressive constructions of mind. *Cognitive Science* **5**, 1–30.

Lewis, M. W. and Anderson, J. R. (1985). Discrimination of operator schemata in problem solving: learning from examples. *Cognitive Psychology* **17**, 26–65.

Matz, M. (1982). Towards a process model for high school algebra errors. In *Intelligent tutoring systems* (eds. D. Sleeman and J. S. Brown), pp. 25–50. Academic Press, New York.

Morales, R. V., Shute, V. J., and Pellegrino, J. W. (1985). Developmental differences in understanding and solving simple word problems. *Cognition and Instruction* **2**, 41–57.

Resnick, L. B. (in press, a). Constructing knowledge in school. In *Development and learning: conflict or congruence?* (eds. L. S. Liben and D. H. Feldman). Erlbaum, Hillsdale, NJ.

Resnick, L. B. (in press, b). The development of mathematical intuition. In *Minnesota symposium on child psychology. Vol. 19* (ed. M. Perlmutter). Erlbaum, Hillsdale, NJ.

Riley, M. S. (1981, January). *Representations and the acquisition of problem-solving skill in basic electricity/electronics.* Paper presented at the Computer-based Instructional Systems and Simulation Meeting, Carnegie-Mellon University, Pittsburgh, PA.

Riley, M. S., Greeno, J. G., and Heller, J. I. (1983). Development of children's problem-solving ability in arithmetic. In *The development of mathematical thinking* (ed. H. P. Ginsburg), pp. 153–96. Academic Press, New York.

Sackur-Grisvard, C. and Leonard, F. (1985). Intermediate cognitive organizations in the process of learning a mathematical concept: the order of positive decimal numbers. *Cognition and Instruction* **2**(2), 157–74.

Sleeman, D. (1982). Assessing aspects of competence in basic algebra. In *Intelligent tutoring systems* (eds. D. Sleeman and J. S. Brown), pp. 185–99. Academic Press, New York.

Sleeman, D. (1984). An attempt to understand students' understanding of basic algebra. *Cognitive Science* **8**, 387–412.

Sweller, J. and Cooper, G. A. (1985). The use of worked examples as a substitute for problem solving in learning algebra. *Cognition and Instruction* **2**(1), 59–89.

VanLehn, K. (1983). On the representation of procedures in repair theory. In *The development of mathematical thinking* (ed. H. P. Ginsburg). Academic Press, New York.

Vergnaud, G. (1982). A classification of cognitive tasks and operations of thought involved in addition and subtraction problems. In *Addition and subtraction: a cognitive perspective* (eds. T. P. Carpenter, J. M. Moser and T. A. Romberg), pp. 39–59. Erlbaum, Hillsdale, NJ.

Vergnaud, G. (1983). Multiplicative structures. In *The development of mathematical thinking* (ed. H. P. Ginsburg), pp. 128–74. Academic Press, New York.

Author index

Subject index